Oil in the Deep South

Oil in the Deep South

A History of the Oil Business in Mississippi, Alabama, and Florida, 1859–1945

Dudley J. Hughes

Published for the Mississippi Geological Society

University Press of Mississippi
JACKSON

A special thank you to Carroll Brinson
for his contributions to this book
and the encouragement he gave to the author.

Copyright © 1993 by the University Press of Mississippi
All rights reserved
Manufactured in the United States of America

96 95 94 93   4 3 2

The paper in this book meets the guidelines for permanence and durability of the
Committee on Production Guidelines for Book Longevity of the Council on Library
Resources.

Library of Congress Cataloging-in-Publication Data

Hughes, Dudley J.
   Oil in the Deep South : a history of the oil business in Mississippi, Alabama,
and Florida : 1859–1945 / Dudley J. Hughes.
      p. cm.
   Includes bibliographical references and index.
   ISBN 0-87805-615-7 (alk. paper)
      1. Petroleum industry and trade—Mississippi—History.
2. Petroleum industry and trade—Alabama—History.   3. Petroleum
industry and trade—Florida—History.   I. Title.
   HD9567.M7H84   1993
   338.2'7282'0975—dc20   92-37759
                                                     CIP

British Library Cataloging-in-Publication data available

Contents

Preface

The discovery of oil in Pennsylvania in 1859 spawned an industry that was to change civilization. The industry spread quickly from Pennsylvania to other states, then to other countries. Similar to early gold rushes, drilling for oil so excited the imagination that a virtual army rushed from one boom site to the next in the early years.

Soon after the turn of the century, oil booms developed in Kansas, Oklahoma, California, and Texas. In 1901 at Beaumont, Texas, the discovery well for Spindletop surpassed all previous records, gushing 100,000 barrels per day and focusing attention on the coastal areas of the Gulf of Mexico. Oil was found in Louisiana in 1902 and in Arkansas in 1921. Natural gas, which in earlier years was a waste to be flared, was harnessed by pipelines to become a major fuel in the twentieth century. This gave rise to a whole new industry and made possible an even higher standard of living.

By the end of the 1930s, the industry over its 75 years had developed systematic methods of exploring for oil and gas and a myriad of geologists and geophysicists were studying the subsurface to locate new places to drill. But in this advanced stage of the industry, no oil had been found in the states bordering the Gulf of Mexico east of the Mississippi River. While the neighboring states of Texas, Louisiana, and Arkansas had become prolific producers with thousands of wells yielding millions of barrels of oil annually, only a few isolated gas fields had been found in Mississippi and northern Alabama. These small gas fields did not inspire much interest. The real profits lay in oil.

Many experts were pessimistic about the probability of any significant oil finds being made in the southeastern states. (One prominent geologist is quoted as saying, "I'll drink all the oil found east of the Mississippi River.") This all changed in 1939 with the discovery of oil in Mississippi at Tinsley. Experience had shown that fields of 100 million barrels were very rare and considered major. Tinsley proved to be a prolific 200-million-barrel oil field and was east of the Mississippi River "barrier."

Coincidentally, a worldwide shortage of oil was developing brought on by World War II. The relatively undrilled southeastern states suddenly appeared as the happy hunting grounds. A drilling boom began in Mississippi in the early 1940s and spilled over into Alabama and Florida in the same decade.

Now another 50 years have passed and oil production in the United States has peaked and is on a fast decline. Most petroleum yet to be found is very deep or offshore or in hostile environments such as the Arctic. Drilling in the old producing areas in the lower 48 states finds mostly very small accumulations and is only worthwhile if petroleum is sold at very high prices.

The oil industry now is on the decline, even in Mississippi. The history of the oil business in almost every sector of the United States except the southeastern states has been recorded. This book documents the history of the petroleum business in Mississippi, Alabama, and Florida, not only by recording a statistical and chronological summary, but also by highlighting the many people and companies who were involved. Even though statistics may forever be available, the personal stories of the companies, entrepreneurs, promoters, investors, and workers who made up the oil business will soon be lost.

My career in the oil business in the southeastern states began in 1953, 14 years after the discovery of Tinsley. Since that time, I have been closely involved in all of the oil and gas plays that have developed in this area. It is with regret that I watch the industry decline. Future generations will never know the great excitement of drilling for oil and gas.

The Oil Business of Today

The search for petroleum is a fascinating business. Once caught in its web, few are ever content with any other lifelong occupation. The basic motivation for becoming involved in the oil business is money, as with any other industry. Oil creates new wealth for the party that finds it and the country in which it is found; it is not the type of wealth that depends on one party's loss for another party's gain, as in trading stocks. Each barrel produced generates original income and adds to the gross national product. Many former Third World countries became international economic giants with the discovery of oil.

It is more than the lure of riches, however, that draws people to this industry. Some find great satisfaction in unraveling the history of the earth. They face the ultimate test, the big gamble of drilling a hole deep into the subsurface to penetrate rocks that have never been seen by human eyes. And they experience a sense of accomplishment in bringing large quantities of petroleum to the surface to heat the homes and fuel the automobiles of our society.

The most romantic, most exciting, and sometimes the most disappointing part of the business is in exploration, the search for petroleum. It involves many types of people: entrepreneurs, geophysicists, geologists, landmen, drillers, roustabouts, financiers, and thousands of helpers. Most of these individuals are constantly on the move from one drilling rig to another from one prospect to another from one state to the next, and, in some cases, from one country to another, wherever finding oil seems most promising. Sometimes the search takes place only on maps, but the intoxication is the same.

Oil exploration is a 24-hour job. During good times, rigs may drill for months without stopping. A shut-down occurs only long enough to move the rig to a new location and start again. Key people are called night and day. The phone rings at 3:00 A.M.:

> "We've twisted the drill pipe off!"
> "We've had a show of oil!"
> "We can buy the lease for $50 per acre!"
> "We will be logging at 6:00 A.M.!"
> "I need the string of pipe delivered by noon!"

The oil searcher's spouse learns to sleep through these nightly interruptions. What stimulates these individuals to such an extent that they can hardly wait to get to the office or field each morning? Why had they rather work on Saturday than play golf? The attraction of exploring for oil becomes almost addictive.

Exploration requires the efforts of a small army of people, working many months, to bring a well into existence. One team locates a prospective drillsite. A second team acquires leases and drilling rights. A third team then drills and equips the well to the point that it produces oil or gas. When the drilling is finished, these people have only mild interest in the completed well. They are instead more interested in the next challenge the next site to be drilled.

Once a well is "brought in" and put on production, it has a life of its own and requires very little attention thereafter. Like a shepherd with a docile herd of sheep, one "pumper" can handle a large number of producing wells. Even offshore platforms with many wells may require no permanent attendant, only an occasional visit by boat or helicopter.

The oil industry consists of three segments: exploration and production (E&P), refining and marketing. Companies engaged in all three segments of the industry are referred to as "integrated" companies. On Wall Street, the largest integrated oil companies are referred to as the "majors," and smaller integrated companies, or those involved in only one or two segments of the industry, are known as "independents." In oil circles, however, "independents" usually mean those individuals, partnerships, or companies that are involved almost exclusively in exploration and production. Oil or gas produced by independents is "sold at the well head" to refining companies and pipelines.

Exploration begins with "wildcat wells" that probe areas where oil has not been found before. Once a wildcat well strikes a new field, "development wells" are drilled in a regular pattern around the discovery until the limits of the pool are established. Wildcat wells have a very high risk of failure, whereas development wells have a reasonably good chance of being successful.

Fortunes can be made or lost overnight in oil exploration. Big sums of money must be risked at a moment's notice to beat out competition in leasing and drilling new wells. It's not a game for the weak hearted.

America's Supremacy

The petroleum industry was spawned in the United States. This country can credit much of its ascension to power to its dominance in oil. For the final 40 years of the nineteenth century and the first half of the twentieth century, the United States produced more oil annually than all the countries in the rest of the world combined. This global superiority continued as late as 1950, when the United States was still producing 52 percent of the world's oil. Its share had fallen to 12 percent by 1990.

The United States has also always been the world's largest consumer of petroleum per capita, far exceeding any other country. Power fueled by oil has contributed greatly to the exceptionally high standard of living enjoyed by the citizens of the country, even though its cheap abundance has encouraged waste of this valuable resource.

In his Pulitzer Prize-winning book *The Prize*, Daniel Yergin emphasized the tremendous importance of petroleum in today's civilization:

Though the modern history of oil begins in the latter half of the nineteenth century, it is the twentieth century that has been completely transformed by the advent of petroleum. . . .

Oil is the world's biggest and most pervasive business, the greatest of the great industries that arose in the last decades of the nineteenth century. . . . No other business so starkly and extremely defines the meaning of risk and reward and the profound impact of chance and fate. . . .

The twentieth century rightly deserves the title "the century of oil."

Perhaps no private enterprise has done more for modern America and its citizens than the petroleum industry.

Introduction

SETTING THE STAGE

In the late 1700s, humans began to increase their ability to produce and transport goods without having to rely on manual labor, harnessed animals, or water or wind power. The new steam engine could generate the power of many horses and led to development of machines of many types.

The fuel to fire the boilers was wood or coal, and lubrication was from animal or vegetable oil. These organic lubricants were high in acid, which ate at the metals and caused machinery to wear out quickly.

In the eighteenth and early nineteenth centuries, tallow candles and oil lamps were used for lighting. By the 1850s, lamps were commonly fueled from four sources: lard oil from pig fat; turpentine from pine trees, which was dangerously volatile; alcohol which emitted only a dull blue light; and the preferred fuel sperm oil. This oil from whales was almost smokeless and odorless and was far superior to other oils (Williamson and Dawn, 1959).

The demand for sperm oil steadily increased in the United States. By the early 1800s, New England whalers had killed off the whales in nearby U.S. waters and were forced to search for them in the Arctic (Williamson and Dawn, 1959). Even in the golden age of whaling, 1830 to 1840, demand for the oil outpaced the supply. In 1850, sperm oil was bringing $2.00 to $2.50 per gallon, amounting to $84.00 to $105.00 per barrel in 1850 dollars (Clark, 1963). In 1860, the wholesale price per gallon for each fuel averaged $0.95 for lard oil, $1.62 for sperm oil, $0.35 for turpentine, and $0.43 for alcohol. The need for a good, less expensive fuel led to the development of "coal-oil."

Coal Oil, Forerunner of Crude Oil

A scientific breakthrough came in 1848 when James Young of Scotland succeeded in recovering oil from coal and shale. He licensed this process to companies in the United States. Young also established a thriving Scottish shale oil industry.

By 1853 Dr. Abram Gessner of Pittsburgh, Pennsylvania, was manufacturing an improved illuminating oil from coal, which he called "kerosene." Nevertheless, "coal oil" became the most commonly used name. As lamps using coal oil were invented and put on the market, coal oil cut deeply into the sperm oil market.

By 1859, there were at least 33 coal oil refineries in the United States, producing 20,000 gallons of coal oil per day. This industry had a short life span, however. Illumination oil from coal disappeared when crude oil from petroleum became plentiful. Nevertheless, coal oil operators developed the essential basis for petroleum distillation and refining, as well as a start toward marketing (Williamson and Davis, 1959).

Manufactured Gas Used for Lighting

Even before the advent of coal oil, "artificial" gas was used for lighting. As early as 1784, gas manufactured from coal was experimentally used for illumination in Europe.

In the United States, the city of Baltimore was using gas to light streets by 1817. Most of the larger cities, eager to follow, established gas works, to light streets, businesses, and even residences: New York in 1823, Boston in 1829, St. Louis in 1847, Washington, D.C. in 1848, Chicago in 1850, and San Francisco in 1852. Gas prices were as high as $10.00 per 1,000 cubic feet, compared to $5.00 in 1990, after 90 years of inflation.

Most of the cities in the southeastern states joined this trend and established city gas works. New Orleans had gas lights by 1832, Memphis by 1852, and Atlanta by 1856. In Mississippi, the Jackson Gas Light Company was incorporated for the purpose of supplying the city of Jackson with gas lighting in 1860, and was in service by 1868. The gas works was situated on the west lawn of the city hall where the statue of Andrew Jackson now stands.

Remnants of the original piping for Jackson's gas light system can sometimes be seen in construction excavations in the oldest portions of the city. The pipes for transporting the artificial gas were bored-out pine logs, with bell joints sealed with burlap and tar. The wooden pipes were slowly replaced by cast iron, probably around the turn of the century. Similar developments took place in almost all southern cities, including Vicksburg, Natchez, Meridian, Hattiesburg, Montgomery, Tuscaloosa, Birmingham, Mobile, and Pensacola.

In the 1880s electric incandescent lights began to compete with gas lights.

Jackson switched to electric lighting in 1888. Manufactured gas was also used for its heating value in stoves and furnaces. Gas lighting of streets was still common in some cities, however, as late as World War II. Eventually, the manufactured coal gas would be replaced by natural gas from gas wells. Most cities used the pipeline distribution networks, already in place for manufactured gas to carry this new fuel.

Private Mineral Ownership

The privilege of private citizens to own mineral rights was probably the most significant single factor to put the United States far ahead of other countries in the development of petroleum resources. In the United States, land-owners own the rights to any minerals found on their land "to the center of the earth." This property right is almost unique to the United States. Private mineral rights made it possible for thousands of companies, large and small, as well as individuals to participate in the frantic search for oil and gas. Drilling rights could be quickly obtained directly from farmers and other landowners.

Even today, in almost every other country, underground minerals belong to the government. To obtain drilling rights, parties must deal with the slow pace and, sometimes corruption of government bureaucrats. Only the largest companies can afford the time and expense. As a result, in most other countries, many fewer wells have been drilled and only very large petroleum accumulations are profitable.

Because of that special freedom to deal directly with individual land-owners, some three million wells have been drilled in the United States, whereas as of 1990 fewer than a million wells had been drilled in all the other countries of the world combined.

Part I

The Development of Petroleum in the Southeastern States to 1925

CHAPTER 1

The Birth of the Oil Business, 1859–1925

Naturally occurring oil from underground sources was found in small quantities at a few locations in the world where it had seeped to the surface. This was often called rock oil or mineral oil. It later became known as crude oil.

One of the earliest efforts to commercially market products distilled from rock oil on a commercial basis in the United States was made in 1858. Samuel M. Kier of Pennsylvania built a five-barrel still to convert "rock oil" into kerosene. Further analysis showed that, in addition to illuminating oil, rock oil also yielded gas, paraffin, and lubricating oils. Kier sold some of the new illuminating oil for as much as $1.00 per gallon while sperm oil was selling for $1.50. During his first year of operation, 1,183 barrels of petroleum were refined. Kier was plagued by limited crude oil supplies, as were others attempting to refine petroleum.

Entrepreneur George H. Bissell, a lawyer and businessman from Connecticut, learned that rock oil had better illuminating qualities than coal oil. The oil came from saltwater springs at Titusville, a town in the northwest corner of Pennsylvania, where Bissell bought a 100-acre plot. In New York, he and others formed the Pennsylvania Rock-Oil Company.

Colonel Drake Drills for Oil

Bissell decided to involve his trusted friend Edwin Lauretine Drake in the project by hiring him to investigate the property and clear up title defects. Later Pennsylvania Rock-Oil Company leased its hundred-acre plot to a new company formed by Drake, called Seneca Oil Company. Drake spent several months attempting to improve the oil springs by digging pits and channels, with minor success. When he tried to dig a well by hand, ground water flooded the shaft and forced abandonment.

Finally, Drake conceived the idea of drilling a well patterned after salt-well drilling, to increase the flow of oil. Saltwater was evaporated to obtain the rock salt which could be sold for a profit. Cast-iron pipe was commonly used to stem the flow of fresh water from shallow sands.

A salt-well driller, William A. Smith, was hired and began drilling using a primitive cable tool rig. The Drake well was the first known well in history to be drilled exclusively in search of oil. On August 27, 1859, while drilled to a depth of 69.5 feet, the hole filled with oil. The well pumped 25 barrels of oil per day and oil was sold for $18 per barrel. Drake's well started a boom that spread like wildfire (Clark, 1955). The Titusville region of western Pennsylvania was inundated by thousands of people. Many wells were drilled in the vicinity of the Drake discovery and soon drilling was extended to other areas.

By the end of 1860, a total of 240 wells had been drilled in Pennsylvania, of which 201 were productive. The same year oil was found in West Virginia, Ohio, Kentucky, and Tennessee. Across the continent, California became a producing state in 1861. The search for oil widened to other states, but Pennsylvania would remain the principal oil-producing state for three decades.

As the Civil War ended, hordes of restless veterans came to the oil fields for jobs. It took thousands of teamsters to transport the oil; carpenters to construct the boomtowns; workers to build roads and railroads; coopers for barrelmaking; derrick builders, drillers, and crews; and workers for all related services.

Oil production in the United States increased from 2,000 barrels in 1859 to 500,000 barrels in 1860, and had climbed to 2,500,000 barrels by 1865. With this flood of oil, the price dropped from $18.00 per barrel in August 1859 to $10.00 by January 1860 and averaged 52 cents in 1861, after falling to a low of 10 cents. Thus, began the rollercoaster rise and fall in oil prices that have been characteristic of the industry ever since. The price rebounded to $14.00 per barrel in July 1864.

Congress, hearing of huge profits being made in oil, and needing money to pay for the Civil War, enacted a tax of $1.00 per barrel in 1865. As crude prices plunged, hundreds of producers were put out of business.

An organized effort by the oil operators finally succeeded in rescinding this tax.

Rapid Progress in Technology and Transportation

Having released this tremendous resource, American ingenuity developed rapidly the equipment and methods to bring it to market. Earthen storage pits gave way to huge tanks made from wooden staves. Transportation by rail, barge, and wagons hauling wooden barrels moved the oil from the well-head to the refinery. Soon railcars equipped with wooden tanks were converted to steel tanks, as were ships. Pipelines carried oil to rail or barge as early as 1865. In 1874, a cross-country oil pipeline began transporting 7,500 barrels of oil a day from the oil regions to Pittsburgh, a distance of 100 miles. Soon other pipelines were constructed to carry ever-increasing quantities of oil greater and greater distances.

Many stills were built in the oil regions for refining crude oil into illuminating oil. Soon these refineries were making three grades of kerosine: "prime white," "standard white," and "straw white," and selling them not only to American markets, but also for export. *Illumination oil from crude oil is spelled "kerosine," as distinguished from "kerosene" from coal. However, most retail sales were made as "kerosene," and it was still referred to as coal oil in rural areas.*

Annual crude oil production rose to 4,800,000 barrels in 1869 and 5,205,000 in 1871. Ten years after the Drake well was drilled, American oil was being exported to almost every country in the world, including those of the Far East. Pennsylvania became the Saudi Arabia of the nineteenth century; and the United States was the world's largest oil exporter for many decades. American kerosine and the "kerosene lamp" penetrated many markets of the civilized world; putting sperm oil and coal oil out of business. The rapid rise of the oil business probably saved the sperm whale from extinction.

The Law of Capture

Oil producers soon found that producing wells at maximum rates brought a flood of oil that dropped off rapidly. The hidden source from which the oil came proved to be limited. After a number of years, the pools usually dried up altogether.

Because oil is a liquid that can flow underground from one lease to another, early wells got a larger share of the oil. The legality of draining a neighbor's oil was soon questioned. In 1889, the Pennsylvania Supreme Court decreed that crude oil is a "fugacious" mineral and that the "rule of capture prevailed." The court's judgment was based on an old English common law by which wild animals fugitive creatures of no fixed place belonged to "he who captures them" (Clark and Halbouty, 1972). This ruling encouraged the rapid drilling of as many wells as possible in close proximity, followed by production at a maximum rate. Much waste resulted from flaring gas and rapid depletion of reservoir energy. Aboveground storage also proved a major problem.

Each new big discovery flooded the market with oil and drove the price down. As the flush production diminished, shortages developed and prices rose. Thus were created "boom or bust" cycles. Not until the 1930s did state governments begin to control the spacing of wells and limit rates of production.

With expanding markets, annual U.S. oil production continued to increase to peak at 30,349,897 barrels in 1882. It declined for three years to 21,858,785 in 1885 before increasing again.

Financial Development of Oil Companies

From its dynamic beginning, oil became big business. Recognizing the potential of this new industry, speculators founded new companies and poured millions of dollars into them. Within five years of Drake's discovery in New York alone, oil companies filed capital certifications totaling $350 million; in Massachusetts $160 million and in Pennsylvania $145 million. Hundreds of new companies were formed, almost all with little experience in petroleum. Although most of these early companies have disappeared, a few were the forerunners of today's giant oil companies. Andrew Carnegie, Jacob Jay Vandergrift, and John D. Rockefeller were among the early players.

Faced with the instability of the business, these early oil companies soon developed the idea of combining production, transportation, refining, and marketing. Cooperatives and interlocking partnerships were in place by the 1870s. The Standard Oil Alliance was started in the early 1870s under John D. Rockefeller's leadership and by 1881 had become a giant combination of companies. This led to the formation of the Standard Oil Trust in 1882.

The trust combined 41 companies with a capitalization of $70 million, which enabled Standard Oil to control prices and transportation in order to squeeze out competition. This play was so successful that the federal government ruled it illegal in 1911. The majority of the nation's major oil companies evolved from the breakup of the Standard Trust, including Exxon, Mobil, Sohio, Marathon, Chevron, and Amoco.

Natural Gas Enters the Picture

Many of the thousands of wells drilled in the search for oil yielded both natural gas and oil, and occasionally drillers found only gas. The gas was considered a waste and was usually blown off in hopes that the well would increase in oil production. In a few cases, natural gas was used near its source as a fuel for small local industries or gas lighting.

Beginning in 1870, an attempt was made to utilize natural gas on a commercial basis. Gas was piped from West Blooming Field to Rochester, New York, 25 miles away, in a pipeline made of Canadian white pine logs bored out eight inches. In 1872 the natural gas was turned into the Rochester distribution mains, designed for manufactured gas. Natural gas, being of higher heating value, did not work well in the system. The wooden pipeline also proved unsatisfactory and the project soon collapsed.

The introduction of wrought-iron pipe with couplings proved to be more successful. In 1891 Indiana Natural Gas & Oil Company laid two parallel lines, eight inches in diameter, a distance of 120 miles from natural gas fields in northern Indiana to Chicago. These operated under a pressure of 525 pounds per square inch. The lines were supplied from an area known as the "Gas Belt of Indiana," which had been discovered in 1886 with wells 900 feet deep. Other cities and industries began tapping this source, so that the supply was rapidly depleted. By 1907 the cities were largely relying again on manufactured gas.

This fledgling natural gas industry, prior to 1900, was most prominent in Pennsylvania, Ohio, and Indiana. In 1882, records indicate a total of $215,000 in sales of natural gas in the U.S., whereas sales had risen to nearly $15,000,000 by 1892. Manufactured gas was still firmly entrenched in the cities at the turn of the century and would remain so for another 30 years. Gradually natural gas would squeeze out the "artificial gas" (as it was often called) as kerosene had put the whalers out of business.

In 1906, 388 billion cubic feet of natural gas were supplied to 879,944 domestic customers at an average rate of 22.7 cents per MCF and 9,074 industrial customers at a rate of 7.8 cents per MCF. By 1923, production had reached 1,008 billion cubic feet sold to 3,232,800 domestic and 18,000 industrial customers at rates averaging 51.1 cents domestic and 13.4 cents industrial (Stotz and Jamison, 1938).

Geologists Join the Team

Geologists played little part in the oil industry during the first decades. Oil seeps, look alike surface features, hunches, and superstition were generally the bases for locating wildcat wells. After a new field was discovered, wells were drilled in every direction until dry holes outlined the limits of the field. The fields became a forest of closely clustered derricks. Drillers became homespun geologists, recording in their local jargon the nature of the beds their bits penetrated. Many states had also geological departments to make surface geology maps and keep records of mineral production within the state. Between 1874 and 1887, the Pennsylvania Geological Survey produced five volumes devoted to oil and gas. Many drillers' well logs were studied, but no correlation was made as to the structural association of oil production.

The anticlinal theory of oil accumulation was suggested as early as 1861, but it was generally discounted until the 1880s. According to this theory, oil had accumulated in anticlines, folds in the earth's sedimentary layers. Trapping occurred when oil "floated" to the top of water that filled porous beds in the anticline. I. C. White, a geologist working for the Pennsylvania Survey, was intrigued with anticlinal theory and resigned to pursue private investigation. In 1885 he issued a report that confirmed the theory so conclusively that it was quickly accepted by other geologists. However, most of these findings were considered academic by the oil industry. Not until after the turn of the century would geologists be commonly employed by oil companies.

In 1916 the American Association of Petroleum Geology (AAPG) was formed with 94 members. This organization had grown to 37,000 members by 1990.

Oil Producing Areas

From early geological surveys, it was learned that Paleozoic rocks 225 million to 600 million years old are found in northeastern and mid-continent portions of the United States. Pressure and temperatures have solidified these rocks into very hard limestones, sandstones, and shales.

In the Appalachian area, millions of years of erosion have removed much of the overburden, leaving these old rocks near the surface that were once buried thousands of feet deeper. For this reason, oil generated at much greater depths could be found at depths of 300 feet or less in the Titusville, Pennsylvania, area. These shallow conditions existed in all of the early producing states to varying degrees, including Pennsylvania, New York, Ohio, West Virginia, Kentucky, Tennessee, Indiana, and Illinois. Paleozoic rocks also outcrop in the extreme northern portion of Mississippi and Alabama. Hence, much of the early drilling in these two states was concentrated in the northern areas.

The surface area of Texas, Louisiana, Mississippi, Alabama, and Florida is made up of sediments once deposited in the Gulf of Mexico, except in the extreme northern portions of Mississippi and Alabama, and in west Texas. The Gulf of Mexico has progressively subsided and has been filled in by sediments since early Mesozoic times. The ancient seabeds become progressively younger and "softer" as approaching the coastline from inland. These more recent deposits are less compacted by time and pressure than the Paleozoic beds in the Appalachian region.

In the Mesozoic age, from 135 million to 225 million years ago, the Gulf of Mexico was much broader. Some layers of rock laid down then are now exposed to form rims around the Gulf several hundred miles inland. Tertiary beds from two million to 65 million years ring the coastline and underlie the waters of the gulf. Before 1895 it was generally accepted that petroleum occurred only in the older, hard Paleozoic rocks and that the soft, younger rocks of the Gulf Coastal Plain had no potential.

Development of the Drilling Rigs

Early drilling during the Titusville boom utilized the cable tool rig. First a wooden derrick was erected from timbers hewn from nearby woods. A hemp rope was then strung over a pulley with a heavy steel chisel like bit attached. A steam engine or manual spring-pole then lifted and dropped the bit, literally chipping a hole in the rock. The churned mixture of water and rock cuttings was bailed from the bottom of the hole. Drilling was very slow, usually no more than five to ten feet per day, and proceeded only during daylight hours. Excessive water in the hole from exposed water sands slowed the rate even further. The bit was effective in hard, brittle beds, but became mired in soft, gummy beds. Drilling depths below 1,000 feet were rare.

By the 1870s many improvements were being made. The hemp rope was replaced with steel cable. Cast iron or steel casing was used to shut off water zones. Drilling to 2,000 or 3,000 feet was common. Even so, drilling was effective only in hard-rock areas. At the time, those in the oil business almost unanimously believed that oil occurred only in hard rocks.

The cable tool rig persisted as the principal means of drilling until 1925, and is still in limited use today.

In Texas, oil was being produced in minor amounts by 1889. However, it was not until 1895 that the first significant field was found and started Texas on its way to becoming the largest oil producing state in America. This happened by a lucky accident.

In Corsicana, near Dallas, a new type of drilling rig, the rotary rig, had been developed for drilling water wells in soft beds. This process involved using a rotating pipe with a homemade "fishtail bit." As the pipe was rotated, mud was circulated down the drill pipe to bring up cuttings as the mud traveled back to the surface on the outside of the pipe. Shallow-water sands presented no problems. In 1895 drillers in Corsicana discovered oil rather than water. By 1900, several hundred wells had been drilled in the oil field using rotary rigs. This was not a prolific field, but it showed that drilling in soft, water-bearing formations was practical and that oil could be present in soft younger rock.

The Gasoline Engine Spurs Demand

Before the turn of the century, the main products derived from petroleum were kerosine, fuel oil, lubricating oil, grease, paraffin, and tar. Kerosine for illuminating purposes was the most desired product. Gasoline was a nuisance to refiners, who considered it a waste to be disposed of, having very limited uses. Gasoline was known by the name, "naphtha."

The development of the gasoline engine in Europe was to make gasoline the choice product derived from oil. By 1885, a satisfactory gasoline engine had been perfected. In Europe, automobiles were being built by 1890. Henry Ford and others were building automobiles in Detroit by 1896. These forces would soon bring a tremendous increase in demand for gasoline.

Petroleum's First Forty Years

As the nineteenth century came to a close, the petroleum industry had completed its infancy and reached an adolescent stage. Almost every rural home in the United States was lit by kerosine lamps.

Oil production in the United States continued to grow. Pennsylvania's annual production was greater than any other state's for the industry's first 35 years. However, production peaked in Pennsylvania in 1891 and declined sharply by 1900. Ohio surpassed Pennsylvania in 1896 to become the dominant producing state. West Virginia was the second most prolific producer by 1900.

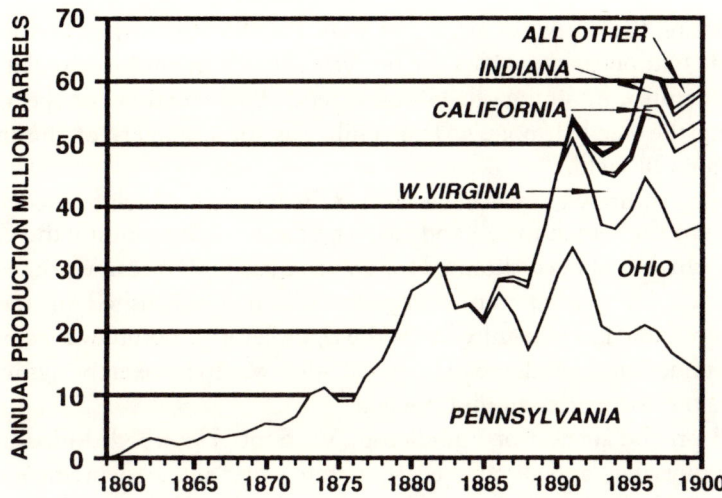

FIG. I   Oil production by states, 1858–1900. The curve representing Pennsylvania
production also includes New York's, which is a relatively small amount.
Production listed as "all other" includes Kentucky, Tennessee, Colorado, Illinois,
Kansas, Texas, Missouri, Oklahoma, and Wyoming.

The last half of the nineteenth century was the "kerosene age." Kerosene
from coal had made inroads into the sperm oil industry only to be displaced
by kerosine derived from petroleum as the principal illuminating oil in the
world. Fuel oil from petroleum was coming into common use, displacing coal
in many furnaces. Petroleum-based asphalt was being used in sidewalks in
some cities.

The Twentieth Century

The twentieth century would be the "golden age" of petroleum as its uses
quickly multiplied. As Detroit turned out a stream of automobiles, the demand for crude oil steadily increased. The airplane, invented by the Wright
Brothers in 1903, became another consumer of gasoline.

By 1911, the demand for gasoline had surpassed that for kerosine. Gasoline became the predominant product to be derived from petroleum as the
gasoline engine came into greater use. The increased demand spurred the
search for petroleum at an ever accelerating pace. Exploration spread to
areas previously considered to have no potential.

In January 1901 a momentous oil discovery astounded the industrial
world. On the flat coastal plain of Texas near Beaumont, the first real
American gusher came roaring in, producing 100,000 barrels of oil per day

far surpassing the output of any other well. Pattillo Higgins, a self-educated geologist, had pursued an idea for ten years that a mound in the flat plain was a structure containing oil. His persistence finally convinced speculators to put up the needed money. The result was the discovery of the famous Spindletop Oil Field.

Captain Anthony Francis Lucas drilled the big well, which opened a great new area to the search for oil and gas. The flat land surrounding the Gulf of Mexico which had been shunned for 40 years suddenly became prime hunting grounds. Lucas was a mining engineer who had established salt mines in south Louisiana and recognized that the Spindletop structure was associated with an underlying salt dome. This association was to have great significance in future oil discoveries around the gulf.

Based on the success of Spindletop, W. Scott Heywood drilled at Jennings, Louisiana, and discovered that state's first oil in September 1901. This oil field also was found in the soft sediments of the Gulf Coast and related to a salt dome.

World War I (1914–1918)

World War I was the first conflict in which petroleum played a major part in determining the outcome. As Lord Curzon said after the armistice of November 1918, "The Allies floated to victory on a wave of oil." Two factors contributed to the importance of oil in the war: Most of the fighting ships had been built or converted to burn fuel oil rather than coal, and automobiles and trucks were used to move troops. The use of aircraft as fighting machines also increased the need for gasoline.

When the war began in 1914, the United States was producing 65 percent of the world's petroleum. This had increased to 70 percent by 1918. The United States provided approximately 80 percent of the petroleum used by the Allies during the war, while supplying the petroleum needs of the U.S. domestic market at the same time (Williamson, 1963).

The petroleum demands of the Allies were handled with great efficiency as experienced oil people were assigned to control the supply. A petroleum engineer, Mark Requa, was named the first energy czar (Yergin, 1991). A national Petroleum War Service Committee was formed made up of heads of several major companies. It was chaired by Alfred C. Bedford (Standard Oil of New Jersey), and other members included E. L. Doheny (Mexican Petroleum Company), E. C. Lutkin (Sinclair Oil & Refining Company), and J. H. Markhen, Jr. (independent).

It is estimated that 133 million barrels of petroleum products were shipped to the Allies during the First World War (Williamson, 1963). This was an enormous volume at the time, though it may seem insignificant by today's consumption.

The Big Producers

Oil production within the United States continued to increase. Production by years were:

YEAR	BARRELS	VALUE
1900	63,620,529	$75,999,313
1905	134,717,580	84,157,399
1910	209,557,248	127,899,688
1915	281,104,104	179,462,890
1920	443,402,000	1,360,000,000
1925	763,743,000	1,284,960,000

The big producers in 1925 were California 232,492,000 barrels, Oklahoma 176,788,000 barrels, Texas 144,648,000 barrels, Arkansas 77,398,000 barrels, Kansas 38,357,000 barrels, and Wyoming 27,173,000 barrels.

The southeastern states were still listed among the non-producing states at the close of 1925.

CHAPTER 2

Early Developments in Alabama through 1925

The first of the southeastern states to be drilled in search of petroleum was Alabama. Since oil had been discovered in the western Appalachian foothills of Pennsylvania, wildcatters followed these same foothills southwestward into West Virginia, Tennessee, and eventually Alabama. Geological publications called attention to Alabama at an early date thanks to its first State Geologist, Dr. Michael Tuomey, who served in that capacity from 1848 until his death in 1857. Much of his work was lost in 1865 when Union forces burned the campus of the University of Alabama, where the state geological survey was located.

During riverboat trips to Mobile from the survey office in Tuscaloosa, Tuomey observed that a large anticline was apparent from beds exposed in the banks of the Tombigbee River in southern Choctaw County. This became known as the Hatchetigbee Anticline. For his report in 1850, Tuomey is given credit for being the first geologist ever to report local deformation in the Gulf Coast Tertiary beds (Owen, 1975).

The anticline became the site of the first recorded well to be drilled in the southeastern states. In 1884 at Bladen Springs in Choctaw County, Alabama, the Trowbridge No. 1 well was drilled to a depth of 1,345 feet on the north end of the Hatchetigbee; the well was dry.

Small amounts of oil were found in Tennessee as early as 1860, but the state was not destined to become a significant producer of oil and gas. Nevertheless, these finds prompted speculation that petroleum might be found even farther south in the hard-rock areas of northeastern Mississippi and northern Alabama. Records are scarce for this period because a drilling well was not required to be registered with the state.

By 1900, a number of dry holes had been drilled in the hard-rock area of northern Alabama. Oil saturated sandstones, "tar sands," in the Hartselle formation were found on the surface through a number of northern Alabama counties including Colbert, Franklin, Lawrence, and Morgan as well as Tishomingo County, Mississippi. The first published reports of these tar sands were in a bulletin written by the Alabama state geologist, E. A. Smith, in 1894 (he served from 1873 to 1927). The occurrence of these tar sands on the surface prompted much of the early drilling, as had the oil seeps in Pennsylvania.

In 1887, the Hartselle Oil & Gas Company No. 1 well was drilled in Morgan County, near Decatur, to a depth of 1,730 feet. A gas show was encountered at 1,093 feet and an oil show at 1,499 feet. During 1888, Colonel J. A. Montgomery-Reed's Gap No. 1 in Blount County was drilled to a depth of 1,935 feet and abandoned. Next, in Madison County, at New Market, the New York-Alabama Oil Company's Morbet No. 1 was drilled to a depth of 1,077 feet in 1890. Several oil shows were reported between 170 feet and 700 feet. In 1892 and 1893, T. H. Allen drilled two wells in Lauderdale County, near Florence: the No. 1 to a depth of 2,750 feet and the No. 2 to a depth of 1,460 feet. The No. 2 well had oil shows from 444 feet to 447 feet and is reported to have bailed oil at four to five barrels per day. The well, however, was abandoned. No doubt other wells were drilled in Alabama before 1900, but there are few records.

Huntsville and Hazel Green Gas Fields

Success came in 1902. In northern Alabama, near Huntsville in Madison County, the New York-Alabama Oil Company discovered natural gas in the Silurian limestone at a depth of 278 to 300 feet. Fourteen wells were drilled, and these supplied the cities of West Huntsville and Huntsville during 1907 and 1908, and possibly for a short period thereafter. The Huntsville Gas Field was the first commercially exploited production in the southeastern states.

New York-Alabama also completed seven gas wells near Hazel Green in Madison County at a depth of 310 to 375 feet. This gas find at such a shallow depth lacked the reserves to sustain production for any long period. Whereas oil does not compress to any extent and may have almost equal reserves at shallow or deep levels, the amount of gas in a reservoir depends on pressure, which becomes greater with depth. Underground pressure increases at an average rate of 0.465 pounds per square inch per foot of depth. Gas at 300 feet would likely have less than 140 pounds per square inch of reservoir pressure. The same reservoir at 5,000 feet should average 2,325 pounds per square inch and have 16 times as much reserves.

Fayette and Jasper Gas Fields

A more significant discovery, the Fayette Gas Field, was made in 1909, in Fayette County north of Tuscaloosa, Alabama. The discovery well, Eureka Oil & Gas No. 1 Providence, produced gas from a depth of 1,400 to 1,420 feet at a rate of 1,600,000 cubic feet of gas per day with a shut-in pressure of 630 pounds per square inch. Additional wells were drilled, including the Providence No. 8 in 1910, which produced at a rate of 4,500,000 cubic feet per day. In 1912, another gas discovery was made near the town of Jasper, Walker County, Alabama, resulting in several gas wells at depths of approximately 1,800 feet.

First Natural Gas Market in Alabama

From the shallow gas discoveries in northern Alabama, the first gas companies were formed to supply local customers. The annual bulletin from the Geological Survey of Alabama, "Statistics of the Mineral Production of Alabama, 1913," indicated there were 340 domestic customers and two industrial customers in the towns of Fayette, Jasper, and West Huntsville, double the number of customers of 1912.

At the close of 1913 there were 18 producing gas wells in Alabama, one producer having been abandoned and seven dry holes having been drilled during the year. By 1914, consumers had increased to 395 domestic and two industrial. Two new producing gas wells were completed during the year with four wells having been depleted. From sketchy records, 16 active gas wells were remaining.

The records of 1915 indicate 291 customers in Fayette being supplied by three wells in the Fayette Field and 150 customers in Jasper being supplied by Jasper Field wells for a total of 441 customers. Apparently, one new gas well was drilled. A decline in customers began in 1916, with 250 consumers in Fayette and 85 in Jasper. During the year, nine gas wells ceased to produce in Madison County, and one in Winston County.

A year later, Alabama gas production had dropped to insignificant levels. Three companies, the Fayette Light and Fuel Company (Fayette), Dixie Gas

Company (Birmingham), and the Jasper Natural Gas Company (Jasper), had distributed the natural gas.

It is not surprising that the early natural gas business in northern Alabama suffered the same fate as many other early natural-gas ventures. When supplies were depleted, the business floundered.

A well near Cardova drilled in December 1916 yielded oil measurable by the barrel, but was not commercial. This stimulated additional drilling in the area, however. By 1917, more than 40 wells had been drilled in the northern Alabama area, mostly around Fayette.

During the next decade, a few wells were drilled each year throughout the state, resulting in some 140 wells with no significant discoveries by 1925 (reported to be 158 by 1927).

S. A. Hobson, the First Alabama Wildcatter

Early drilling in Alabama had been brought about by men from other oil areas. The Fayette discovery involved the first native son to become an independent oil wildcatter in the southeastern states, S. A. Hobson. Hobson was the brother of the famous admiral, Richmond Pearson Hobson, who is remembered for his naval gallantry during the Spanish-American War.

S. A. Hobson's 1920 publication, "Ups and Downs in Old Alabama (From the Oil and Gas Standpoint)," sets out his version of the first commercial production in Alabama:

> I drifted along like flotsam on the beach for about a year, when I succeeded in interesting Mr. Brennan in a coal land proposition in Fayette and Tuscaloosa Counties, and made with him a tentative agreement that he would drill three diamond drill holes for a certain interest in a certain acreage after I should procure the options thereon.
>
> Armed with this agreement I succeeded in interesting Mr. F. H. Crockard and Dr. G. B. Crowe enough to induce them to put up the funds for a certain acreage. With said funds I secured a much larger acreage than I had hoped to get up. Mr. Brennan drilled one hole and decided to retire. Upon his urgent request, we released him from his contract to drill the other two holes. His cores, however, indicated to me that we were either on the edge, or in the very midst, of a real oil and gas field. Dr. Crowe had great confidence in my judgment on such matters and induces Mr. Crockard to continue with him and myself in the securing of a still larger acreage.
>
> Mr. Crockard, Dr. Crowe, and I let the Providence Oil and Gas Co. have a lease on 500 acres. I had no connection at that time with this

company except as a stockholder. They sank a dry hole and went broke. I still urged that the field was there. The Eureka Co. was organized; I located a hole 300 yards from the Providence hole; it came in like a hurricane on December 20th, 1909.

As is to be expected in such cases, a great bunch of impostors quickly got upon the scene, each one claiming the credit for opening up the field. I stood it as long as I could and then published an open letter giving the real facts, closing with this rather pathetic plaint:

To conclude: I dug up this field. I first put in operation the forces that developed it. I located the first well. This put Fayette County on the map of the world in box car letters. This doubled the value of real estate, and will redouble it again and again. I have never in any way deceived any citizen of the County, and always instructed my agents never to do so. I have been a quiet, respected citizen. I have done my work modestly, and submitted while others took or got the credit. I don't propose any longer to let cuckoos crowd me out of the nest that I built with so much time, labor, study, care, and suffering. . . .

The Eureka Oil & Gas Co., which opened up the field, was incorporated into the previously unsuccessful Providence Co., on the ground of the latter Company's larger capitalization, by an exchange of stock arrangement.

My connection with the developing Company was severed after I had located in rapid-fire succession four more good gas wells, and had indicated the location of No. 15, the last that ever came in. The fact that none other ever came in is due to subsequent bad locations. . . . I had taken the precaution to sell a bunch of my stock (Oh that I had sold it all while selling was good), and with the proceeds acquired an enormous acreage in Fayette County and several other counties of Northwest Alabama, expecting, as a natural consequence of the striking of pay gas at Fayette, and subsequently Jasper and Haleyville, that the oil and gas development of Northwest Alabama would proceed apace. As time began to reel off without results, I grew restive at the delay, as I well might, for my money was all gone for leases which were growing shorter each day. So I decided to become promoter and do some drilling myself.

Hobston published the following article in the Birmingham *Ledger*, on October 13th, 1912:

Congratulations to the State of Alabama and hats off to Mr. Shannon on the bringing in of his third gas well in the Jasper field. Congratulations, too, on the quiet business-like method, so devoid of buncombe and fustian, that has characterized his course. Continued success will crowd such efforts; and I make bold to predict that he will, in the next twelve months, be able to add to his present output (which is ample for all present needs) enough new wells to assure an adequate, permanent reserve for the future growth of the promising town of Jasper. . . .

Mr. Editor, at the top of the superb Alston skyscraper in the fine little city of Tuscaloosa, a huge electric sign displays in star-challenging letters the inspiring injunction: 'Try Tuscaloosa.' And I will say, in conclusion, to the adventurous oil and gas prospector: Try Tuscaloosa, and don't stop till you test the Hartselle sandstone; and by all means don't fail to try Marion County and don't stop short there of the Dolomite.

A further communication from S. A. Hobson appeared in the Sunday Birmingham *Age-Herald* of October 20, 1912:

Open Oil and Gas Fields Right
By S. A. HOBSON

We have been amazingly lucky here in Alabama. I had the honor of starting the fireworks. . . .

When, Mr. Editor, on that crisp December morning in the, to me, epochal year of 1909, I stood in the Sipsey swamp near Fayette, and heard issue from the earth the weird sound that made me know that I had located the first great gas well in our great state, you may well know that I enjoyed the supreme sensation of a life time. Though, in reality, it was the most excruciating discord, about like unto the combined caterwauling of a hundred choruses of underground wild cats; yet, it seemed to me the most delicious music that ever ravished mortal ears.

Hobson was born in 1868 and died in 1962 at age 94. He was actively drilling wells in Alabama well into the 1940s but apparently never had a commercial discovery after Fayette.

Amerada is Founded

The Amerada Petroleum Company had its beginning as the Alabama Exploration Syndicate in 1916. It was the brainchild of Everette Lee DeGolyer, who formed the syndicate for S. Pearson and Son, a prominent British company based in London. DeGolyer's boss was one of the world's leading engineer-geologists, Sir Weetman Pearson, who had been granted the title of Lord Cowdray by the King of England in honor of his accomplishments. Before undertaking his venture in oil, Pearson had achieved world fame for his engineering and financial achievements, dams in Egypt, the Hudson

Tunnel in New York City, a railroad across Mexico's Isthmus of Tehuante-pec, construction of the Port of Veracruz, and many other major projects. By 1900 he was one of the world's richest financiers.

After the oil discovery at Spindletop in Texas (1901) Pearson became enamored with the oil business and made a personal tour of Oklahoma oil discoveries. He then formed the Mexican Eagle Oil Company, which oper-ated in Mexico. The company suffered severe losses in its early years of operations, and "the great Pearson financial empire risked because of the turn of events on finding oil in Mexico was threatened with total doom." (Tinkle, 1970). The company was saved with the discovery of two giant fields, Potrero del Llano and Los Noranjos in Mexico's "Golden Lane" in the Tampico region. Between 1914 and 1919, Mexican Eagle became pos-sibly the richest oil enterprise in the world.

One of the early employees of Mexican Eagle, Everette Lee DeGolyer soon became its leading geologist. However, the business climate in Mexico eventually declined with a revolutionary change in government, causing Pearson to extend his oil operations into the United States. A portion of Mexican Eagle was sold in 1919 to Royal Dutch Shell, which took over as operator of the Mexican oil interest.

DeGolyer had predicted in 1916 that Texas, Oklahoma, and possibly Louisiana, Kentucky, and Alabama were "the great possibilities of the im-mediate future" for oil exploration. Along with Mexican Eagle's top officials, he organized the Alabama Exploration Syndicate in December 1916 in New York City to give Mexican Eagle a foothold in the United States. Leasing soon commenced in Alabama.

The Alabama Exploration Syndicate became the pilot company for Amer-ada Petroleum Company, which was established in 1919. The first president of Amerada was Thomas Ryder (past president of Mexican Eagle). Everette DeGolyer was named vice president and general manager. He determined that the first operation of the new company would be in Alabama, Texas, Louisiana, and Oklahoma:

> Characterizing Alabama as a very attractive wildcat area, he notes that Alabama Exploration Syndicate has 30,000 acres of leasehold there, where "anticlines are only sixty miles distance from tide water at Mobile." (Tinkle, 1970)

Amerada was almost an immediate success, showing a net profit of some $4 million by 1924, mainly from its oil finds in Oklahoma. The company did not drill its Alabama leases.

DeGolyer would be involved in the petroleum development of the south-eastern states periodically until his death in 1957. Daniel Yergin paid tribute to him in *The Prize*:

> No man more singularly embodied the American oil industry and its far-flung development in the first half of the twentieth century than DeGolyer. Geologist the most eminent of his day entrepreneur, innovator, scholar, he had touched almost every aspect of significance in the industry.

CHAPTER 3

Oil Drilling in Mississippi through 1925

Mississippi had state geologists as early as 1850, but the one who achieved national fame was E. W. Hilgard, appointed in 1858. His famous report of 1860, "The Geology and Agriculture of Mississippi," outlined the geology of the state and reported the existence of the "Jackson Dome" where gas was discovered 70 years later. Frank Smith offers the following profile:

> Eugene W. Hilgard was born on January 5, 1833, at Zweibrucken, Bavaria. . . . where his father [Theodore] was Chief Justice of the Bavarian Court of Appeals. Theodore Hilgard was one of the new liberals who lost their place in the reaction after the new nationalism that followed the defeat of Napoleon.
>
> Theodore Hilgard left Germany in 1835 and settled with his family the next year in Belleville, Illinois, part of a substantial colony of Bavarian refugees.
>
> Young Eugene received a thorough education at home, and Eugene returned to Germany to gain his Ph.D summa cum laude in chemistry, physics, and geology at the University of Heidelberg. . . .
>
> Hilgard went to work for the Smithsonian Institution in Washington, but shortly after reaching there learned that Mississippi was seeking a state geologist. He took the post at once. Working from a laboratory at the University, he made a study of Mississippi soil, and published the results in 1860 under the title "Report on the Soil and Agriculture of the State of Mississippi.". Hilgard's report has become a classic for the state of Mississippi and a guide for similar studies in the rest of the country.
>
> Hilgard celebrated his report with a trip back to Spain to marry the girl

he left behind, Jesusa Alexandrina Bello. The couple came back to Ole Miss to be entrusted with the safekeeping of the school's geological equipment. His primary responsibility during the next five years was to find a way to exist [during the Civil war]. When the school was reestablished in 1866, he took on the duties of Professor of Chemistry.

In struggling post-war Mississippi, Hilgard began to champion the cause of scientific agriculture. . . . When plans were made to establish a separate agricultural school (Mississippi A & M), Professor Hilgard was bitterly disappointed. . . .

Hilgard accepted a faculty position at the University of Michigan, but soon . . . went to the University of California in 1874. . . .

He retired in 1906, but continued as an active Professor Emeritus until his death in 1916.

Initial Attempts to Find Oil in Mississippi

There is no record of wells being drilled for oil in Mississippi until after the turn of the twentieth century. The first was drilled in Clarke County, one-quarter mile east of the railroad station at Enterprise, in the spring of 1903. It was financed by local businessmen who formed the Alabama-Mississippi Investment and Development Company of Enterprise, Mississippi. The well was abandoned at a depth of 1,842 feet, but the same company went on to drill a second well in Clarke County near Barnett in 1905, to a depth of 1,400 feet. The same year, a well was drilled in Tishomingo County near Eastport by a "Captain Brady" to a depth of some 750 feet.

No further drilling in the state ensued until 1911, when three attempts were made. A few wells, all dry, were drilled nearly every year thereafter through 1925.

Pioneer Charles Robert Ridgway, Jr.

One of the earliest Mississippians to become involved in the oil business was C. R. Ridgway, Jr. Ridgway had graduated from Ole Miss Law School in 1908 and established a law practice in Jackson. A few years later, he became intrigued by the oil business to the extent that he passed the bar exam in Texas with plans to move there. His wife refused, however. She wanted their children raised as Mississippians. C. R. capitulated. Instead of leaving the state, he spent a great deal of effort during his lifetime attempting to bring the oil business to Mississippi.

In 1916, he was responsible for the formation of the Jackson Oil and Gas Company, with some of Jackson's most prominent citizens as stockholders, including R. E. Kennington, R. W. Bullard, W. E. Williams, W. G. Stevens, T. M. Hederman, J. A. Gordon, P. F. Culley, A. R. Addkison, Julius Crisler, J. H. Wells, W. E. McGehee, Joseph Ascher, J. M. Hartfield, H. V. Watkins, W. T. Pate, I. Dreyfus, and Ridgway.

The company used its capital to buy leases throughout Mississippi and Alabama. It also participated with minor interest in some wells. However, after several years with no success the stock was redeemed and a new company, the LA-MISS-ALA Oil Company was formed in 1921. This company raised capital of $49,200.

The charter of the new company specifically stated that it would attempt to develop the prospective oil and gas field on the Jackson Dome (pointed out in the Hilgard report). LA-MISS-ALA applied for a franchise to supply natural gas to the city of Jackson, asking voters to approve the franchise before wells were drilled. Apparently, the company was unsuccessful in obtaining voter approval and no well was drilled.

The company initially held 60,753 acres of leases in 16 Mississippi counties with a total investment of $12,631.72. After several years, with no success, the stock of the company was redeemed and The Mississippi Company was formed to deal in real estate, with Robert Ridgway as president, W. E. McGehee as secretary, and W. S. Ridgway, as treasurer.

The Ridgway group would be involved in a number of wells in the Jackson Gas Field in the 1930s, and C. R. Ridgway, Jr., pursued his interest in oil and gas until his death on February 22, 1941. He lived to see the oil discovery at Tinsley, vindicating his lifelong conviction that oil would be found in Mississippi.

The Legendary B. B. Jones

The first native-born Mississippian known to become "rich" from the oil business was Bernard Bryan Jones, known as "B. B." Born in 1866, he completed high school at Kosciusko and attended Mississippi Agricultural and Mechanical College for two years before departing to work for the railroad. In 1907, he left his railroad job, moving to Oklahoma (then Indian Territory) to work with his brother, Montfort. Montfort had opened a bank in Bristow, Oklahoma. As field man for the bank, B. B. Jones was involved in land trading, which enabled him to befriend many of the landowners and Indian chiefs.

When oil and gas came to Oklahoma he had an advantage in buying leases and royalty because of his many contacts. The brothers began to finance Tom Slick's wildcat ventures, with no success at first. In 1912, however, Slick made a location at Drumright, Oklahoma, which proved to be a major discovery and was later known as the Cushing Field.

B. B. and his brother Montfort went on to organize the Bermont Oil Company. By the early 1920s they were immensely wealthy by standards of the day. The brothers moved to two farms in northwestern Virginia on the outskirts of Washington, D.C., where they raised horses. B. B. became a very generous philanthropist, primarily furnishing money to provide education for young people. His benevolence became almost a full-time job. In November, 1919, he formed the Feild Co-operative Association, Inc., for charitable, educational, and scientific purposes in Jackson, Mississippi; it was named in honor of his mother, whose maiden name was Feild. In 1925, a revolving, permanent student-loan fund was established in Jackson, which operates to this day. Thousands of students have been aided by this fund and allowed to complete their college education. Gifts were also made to many colleges, hospitals, libraries, and for other educational and charitable purposes.

Dealmaker T. B. Slick

Tom B. Slick was one of the nation's leading oil men in the first two decades of the twentieth century. Although from Oklahoma City and never a permanent resident of Mississippi, he often traveled to the state, influenced by his good friend and business associate B. B. Jones.

In 1916, Slick was traveling by rail on the Yazoo and Mississippi Valley Railroad to New Orleans via Jackson. The route of the train passed through the flat delta farmlands south of Yazoo City. Abruptly the tracks turned eastward into the Perry Creek Canyon. This valley is flanked by wooded loess hills steeply rising 150 feet on either side of the track. Three miles along Perry Creek, the train passed the Tinsley Switch. Tom Slick, "who had a nose for oil," was struck with the conviction that oil lay under the land he was crossing (Miller, 1990).

On reaching Jackson he left the train. He hired a land agent, and purchased the Valley Plantation, comprising 2,300 acres, in consideration of $12,740.30 cash plus the assumption of two deeds of trust. *Twenty-four years later a portion of this land would prove to be in the heart of the Tinsley Oil Field.*

On March 8, 1927, Slick signed a note to B. B. Jones for $50,000, plus interest at 6 percent, payable on or before two years from that date, at the First National Bank of Oklahoma City. Two receipts for the interest, $1,500 each, were on file before the money was paid. This money was contributed to the Feild Co-operative Association (Helen Youngblood Coffee).

Other Significant Events

Early in the twentieth century several developments in neighboring states would prove of lasting significance to the Mississippi petroleum industry. In 1909, Standard Oil Company commenced the construction of an oil refinery at Baton Rouge, Louisiana. This refinery eventually would be a major buyer of much of the crude oil produced in the southeastern states. In the early years, however, the refinery was supplied by a crude oil pipeline from Oklahoma's Glenn Pool Field. *By 1990, it would be the second largest refinery in the United States.*

A major gas field was found at Monroe, Louisiana, in 1916. Eventually, the field covered 425 square miles with a reserve of seven trillion cubic feet of gas. Only 60 miles from the Mississippi border, it was the largest known gas field in the world at the time. With the limited local markets for this huge gas reserve, Monroe became center of production for carbon black, which was used in the manufacture of rubber, printing ink, stove polish, phonograph records, and other products.

The first carbon-black plant was opened in 1917, but by 1921 there were nine. Six of these were operated in conjunction with plants stripping gasoline from the gas. Lack of market kept the profitability of the field at a low level during its early years (Franks, 1982). It was to become the supplier of the first major gas pipelines that would cross into the southeastern states.

In 1923, Mississippi Power & Light joined two other utilities in the generation of electricity fueled by Monroe gas. Louisiana Power & Light, operator of the project, built a plant at Sterlington on the Quachita River to generate 25,000 kw. A power line was built to Vicksburg to tie into the MP&L System. Low-cost abundant fuel at Monroe made this feasible. This electric power reached Jackson by 1925.

Arkansas still had no significant petroleum production twenty years after the Jennings discovery in Louisiana, even though many more oil and gas fields had been developed in Louisiana during the period. In 1921, a roaring gusher brought in the El Dorado Field and quickly catapulted Arkansas into the column of important producing states. Again the eastward march of petroleum drew attention to Mississippi as a potential producer.

Mississippi—All Dry Holes

Before the close of World War I, some 20 wells had been drilled in Mississippi in search of oil and gas. All were dry. During the 1920s exploration picked up remarkably. By the end of 1925 some 53 wells had been drilled in the state with no success. Most of these early tests were drilled to a depth of less than 2,000 feet with only three or four reaching 4,000 feet.

The oil companies became skeptical of finding any significant production in the state. Even the state geologist expressed doubt. In the Mississippi Geological Survey Bulletin No. 20 (1925) E. N. Lowe stated:

> . . . speaking generally of the prospects of discovering oil and gas in Mississippi, it may be said that thus far the Saline Dome [salt dome] structures with which oil and gas have been found to be so frequently associated in the Beaumont and Jennings oil fields, and in the coastal regions of Texas and Louisiana generally, have not been discovered in Mississippi, although they have been diligently looked for. If they exist here at all, a very detailed examination will be necessary to find them. Of course, not finding them makes a search for oil and gas in south Mississippi less favorable owing to excess depth.

CHAPTER 4

Oil Drilling in Florida through 1925

The state of Florida established two short-lived early posts for state geologist. In 1852 "General" Francis L. Dancy, a former military officer and engineer from St. Augustine, was appointed "State Engineer and Geologist." His principal responsibility was to drain lowlands for agricultural development. His post was abolished three years later by the legislature, after he had asked for $500.00 to do soil tests in various parts of the state. The second state geologist was appointed by Governor E. A. Perry in 1886, prompted by the discovery of valuable phosphate deposits in the state. The post was filled by Dr. John Kost, a "medical doctor and amateur geologist," but the post was abolished in 1887 by the legislature. It was not until 1907 that legislation was passed establishing a permanent Florida Geological Survey.

The first state geologist with the proper credentials and training was E. H. Sellards from Kansas. He occupied this position for 12 years, accomplishing much in ground water studies, paleontology, and Florida geology. In 1919 Sellards left Florida to become a geologist in Texas, where he eventually achieved great professional fame.

Then came one of the most outstanding state geologists of the century, Herman Gunter. A graduate of Florida State, Gunter joined the Florida Geological Survey staff in 1907, when the agency was in its first year. After he became state geologist in 1919, he made the survey a highly accessible source of information on Florida geology.

> His interest in preservation of the water resources of Florida also propelled him to the forefront as an opponent of the Cross Florida Barge Canal,

originally conceived as a sea level ship canal across Florida (Schmidt, 1990).

Gunter was interested in mineral development in the state. He often spoke to geological societies in oil-producing states, and many oil publications carried his comments on Florida. He retired in 1958 after 39 years of service to the state (Florida Geological Survey, Walter Schmidt, 1990).

Drilling Before 1926

Petroleum exploration in Florida was practically independent of developments in the other states. . . . There were no significant geological leads from outside; no showings of hydrocarbons were present; outcrops were sufficient only to indicate broad regional structure. (Pressler, 1947)

Florida records indicate that two wildcat wells were drilled in Gadsden County, west of Tallahassee, in 1892, by the Owl Commercial Oil Company (a subsidiary of the White Owl Cigar Company). The first reached a depth of some 1,000 feet and the second 1,750 feet. Twenty-seven wildcat wells, many of which were too shallow to be important, were drilled between 1901 and 1925. The deepest well was drilled in Washington County, north of Panama City, to a depth of 4,912 feet in 1921.

CHAPTER 5

The Ancient Basins

By the 1920s geologists had established the fact that petroleum occurs in ancient seabeds, known as "basins," which have been filled in by layers of sedimentary rock over millions of years. The sedimentary rock came from mud, silt, and sand being washed into the early seas by streams, by precipitation of carbonates (limestones) from sea water, or in some cases by salts deposited from evaporated sea water. Geologists learned that oil is the altered remains of organisms. Some once lived in the ancient seas, while other organic material came from land plants that was washed into the seas by streams.

The southeastern states were fortunate to have, in their subsurface, a portion of two organically rich basins. The first of these, in northern Mississippi and Alabama, was a depression in the very old Paleozoic sea which covered most of the Appalachian and mid-continental states. This sub-basin came to be known as the "Black Warrior Basin," named after the Black Warrior River in Alabama. Sediments in this basin are from 275 million to 500 million years in age.

To the south, the Gulf of Mexico has been filling in for the past 180 million years, since early Mesozoic times, to form the Gulf Coast Basin. Southern Mississippi, southern Alabama, and all of Florida have thick subsurface layers of sediments deposited in the Gulf Coast Basin. In the western portion of this basin occur many of the prolific oil fields of Louisiana and Texas.

Geologists postulated that both of these basins had the potential for accumulation of petroleum in the southeastern states. In this vast three-state area, petroleum could occur anywhere, and geologists looked for clues as to where to begin their search. Two prominent giant regional structures were known from published reports of the state geological surveys and the United States Geological Survey. These were the Jackson Dome in Mississippi and the Hatchetigbee Anticline in Alabama. Both became early focal points.

CHAPTER 6

The Exploration Teams

By the 1920s oil companies were planning exploration scientifically with large staffs of geologists, geophysicists, scouts, and landmen. Foremost of the group were the geologists, who gleaned information from every source to locate potential new prospects to drill. The American Association of Petroleum Geologists' membership had grown to 1,504 by 1926.

Geologists used well records to study the subsurface strata and predict areas where oil or gas was likely to occur. (Gone were the days of the witching stick.) They also painstakingly mapped beds exposed on the surface, using planetables and transits, searching for indications of underlying structure. Formations normally outcrop in a tilted "layer cake" fashion in predictable bands. A large structure may cause a bed to be out of place as an "inlier" among younger beds on the surface. These inliers led to the discovery of many of the early large fields.

Geophysicists

Geophysics was a newer tool used by the oil companies to locate structures within these basins. The large volume of salt in salt domes has less density than the surrounding beds and actually exhibits a minutely lesser pull of gravity. An igneous intrusion or basement high, on the other hand, actually has a greater pull of gravity.

An early gravity-measuring device was the "torsion balance." A grid of measurements revealed gravity minimums or maximums, either of which could indicate buried structures. Another early geophysical tool was the magnetometer, which measured the strength of the magnetic field at any point on the surface of the earth. When a grid of readings was taken, contours often revealed evidence of deep-lying structures that might be trapping petroleum.

"Refraction" seismograph was being used on the Gulf Coast to locate salt domes by 1925. A more advanced method, "reflection" seismograph was under intense investigation. As the name implies, explosive charges were being recorded from sound waves reflected from hard layers back to the surface. By 1925 the Amerada Petroleum Corporation was developing technology to measure accurately the split-second travel time of the sound waves so that the depth to the reflecting layers could be computed. By shooting a grid of such points over an area, geophysicists attempted to construct a map that could show potential traps for petroleum.

Landmen and Scouts

Landmen had become essential to oil companies. These men dealt with landowners to buy oil and gas leases in areas thought to contain petroleum. In many cases, they bought leases at random, so that in the event of discoveries by other operators, some of the random leases might prove productive. This was called "checkerboarding." When a company's scouts reported that another party was leasing in an area, landmen often rushed in to buy a few leases for their company to capitalize on any drilling success by their competitors.

Oil scouts were essential members of the oil-exploration teams. They kept surveillance on all drilling and leasing by other companies. Scouts also furnished geologists with detailed records of all wells drilled.

Eyes on the Southeastern States

The oil business was 65 years old in 1925. By that time, companies frequently speculated on the oil and gas possibilities in the southeastern states. The progression of discoveries from Texas to Louisiana and on to Arkansas made Mississippi the obvious next target. Interest was to intensify when Mississippi had its first discovery in 1926.

PART II

THE 1926 – 1938 ERA

CHAPTER 7

The Landowners

In the 1920s the southern states were still largely rural. There were very few paved highways in Mississippi, Alabama, or Florida. Winding dirt roads connected the small cities and towns. The principal means of transportation between cities was by passenger train. Mississippi had less than one hundred miles of paved roadway throughout the state. The state paid farmers to maintain the roads adjacent to their land with mule teams. Most of the larger towns had brick streets in the downtown area.

The wealth in Mississippi existed mostly in the rich delta plantations. The Delta of Mississippi is not at the mouth of the Mississippi River, as the term implies. Rather, it is the wide flat floodplain between Vicksburg and Memphis, west of the river, which reaches a width of some one hundred miles near Greenwood. Those families who owned large farms along this fertile floodplain were the aristocracy of the state.

Newer wealth came from lumber. Large tracts of land with virgin southern pine forests had been logged clean, and most of the lumber exported to overseas markets. Although these denuded timberlands now appeared almost worthless, they were taxed at a fixed rate per acre by the state. Many of the lumber companies sold the surface for a dollar or less per acre but retained the mineral rights. Other companies let the land go to the state to be auctioned for delinquent taxes, but often severed the mineral rights first. The larger lumber companies ended up with hundreds of thousands of acres of land or mineral rights in large blocks. When the oil companies arrived, these worthless lands suddenly became valuable for oil and gas leasing. Much of the land in Alabama and Florida was also owned by lumber companies.

In Alabama new wealth was being generated in Birmingham from steel mills. Outcrops of coal and iron ore in close proximity gave rise to this booming industry, which began after the Civil War. Florida had developed a thriving tourist trade along its east coast. Many large luxury hotels were built to accommodate wealthy guests from New York and other northeastern cities during the winter months. These guests presented a sharp contrast to the more impoverished Florida natives. In any case, most of the people of all three states were poor, especially by today's standards.

The business code within the small cities was nevertheless quite formal. Suits with ties and hats were standard attire. No businessman would board a train or attend a meeting without being properly dressed.

Most of what was known about the oil business came from newspaper accounts. Radios were just coming into common use in the 1920s. Stories of oil being found in Louisiana and Arkansas gave rise to hopes that some day it might be found farther east. Previously, oil had been drilled almost exclusively by small promoters, who often asked for free leases. These leases were usually broken into many small lots and sold to speculators until enough money could be raised to drill a well. Most wells were drilled to shallow depths with no geological or geophysical information to improve the odds of finding oil or gas.

Things were about to change. The oil business began serious exploration in the southeastern states in 1926. These states, lying east of the oil plays in the Louisiana and Texas Gulf Coasts, and south of the Arkansas and midcontinental oil fields, had been largely avoided by most of the oil companies. The drilling activity by small promoters depended on money raised from local citizens; rarely was any assistance received from established oil companies.

Many geologists felt that the prospects of finding significant petroleum accumulations in the Gulf coastal states east of the Mississippi River were remote. After all, nothing had been found in all the earlier years of drilling. Consequently, it is not surprising that the first discovery in Mississippi was drilled by local businessmen.

CHAPTER 8

Amory Gas Field

It was 65 years after Colonel Drake's well was drilled in Pennsylvania that Mississippi had its first strike. Gas was found in the hard Paleozoic rocks of the Black Warrior Basin in the northeastern portion of the state.

Citizens of Monroe County speculated that oil or gas might be located in the vicinity of Amory or in adjoining counties. One of these citizens, Charles L. Tubb of Amory, convinced several prominent Mississippi businessmen to invest in a new company to explore the oil and gas possibilities of the area. The Amory Petroleum Company was organized in 1926. Other investors included L. E. Puckett and J. N. Mullins of Amory; Judges W. M. Cox and G. Gillespie of Baldwin; A. W. Reynolds of Starkville; and George L. Shelton of Greenville. Puckett was designated the first president of the company, although Gillespie appears to have taken over when drilling was started.

The new company engaged a geologist, H. D. Easton of Shreveport, to study the surface geology of the territory and recommend a location for the company's first well. Easton chose a site on the "Uncle Bonnie Carter" property, three miles south and six miles east of Amory.

J. P. Evans, a lease broker from Shreveport, was hired to assemble a lease block around the drillsite, comprising some 36,000 acres. The Amory Petroleum Company Carter Oil No. 1 was spudded February 15, 1926, with a cable tool rig. It took almost eight months of daylight drilling to reach 2,400 feet. Finally on October 5, 1926, at a depth of 2,404 feet, the well began to blow gas from the open hole. Drilling was continued to a total depth of 2,412 feet and the flow of gas became a deafening roar. The volume of the original flow was estimated at 5 million cubic feet per day. The original surface pressure was reported at 680 pounds per square inch.

In order to establish a market, it would be necessary to prove the well would hold up on prolonged production. Bill Steinhoff, the driller, added additional joints to the casing to extend the pipe a safe distance above the top of the derrick. The valve was opened, and for 93 days the well was allowed to flow wide open through the 6 5/8-inch casing.

> The gas was fired at the top of the derrick, and for several weeks burned like a great torch, visible for many miles. The total gas thus wasted was estimated at 500,000,000 to 540,000,000 cubic feet. (Vestal, 1943)

As the huge flare roared above the derrick, it beckoned oil men to the scene like moths to a candle. Scouts from many companies rushed to the area and overflowed the Park Hotel at Amory, which was not prepared for the onslaught. The scouts reported back to their companies and soon company executives, geologists, landmen, and independents came to survey the situation.

One of the larger independent operators from out of state, Benedum-Trees, offered Amory Petroleum half a million dollars for one-half interest in the Carter well and the 36,000-acre block with a further consideration that Benedum-Trees would drill Amory Petroleum nine free wells. The Steffey Report of July 25, 1930, states that

> as the original trustees were all inexperienced in oil and gas with a headstrong ex-circuit attorney of St. Louis, Judge G. Gillespie, whose contact with oil was at Smackover [Arkansas] where he ran a hamburger stand, . . . the deal was turned down.

This would prove to be a costly mistake for the Amory group.

With the well flaring wide open, crowds of people came from all over to see the sight. When a gas well is allowed to blow wildly into the air, it creates a deafening roar. The bedlam has been compared to standing under a railroad trestle with a fast train passing overhead, or being next to a jet airplane with all engines revved to full speed. S. A. Hobson called it "the most excruciating discord, about like unto the combined caterwaulings of one hundred choruses of underground wild cats." The ground trembles under this great release of energy, as if one were standing at the epicenter of an earthquake. As the giant flare roared into the heavens, visitors to the Amory well must have quaked inwardly and gasped in awe at the tremendous force of nature that had been released. Could this power underlie the peaceful farmlands of northern Mississippi?

Two of Amory's more colorful citizens were Sam Grady and Hoss-head Smith. Grady ran a livery stable and was at times justice of the peace, town mayor, or sheriff. Smith ran a blacksmith shop and worked on the railroad:

> Sam and Hoss-head had bought a couple of acres of royalty in the area and were anticipating with great relish the certain riches to come. They were driving out to the well one afternoon with a jug between them on the seat of the car.
>
> They were well into the jug when Hoss-head said, "Sam, with all these folks out here every day we ought to put up a stand and sell hamburgers and cold drinks. We could make a lot of money." Sam said, "Awe Hell, Hoss-head, that's a good idea but let's let some poor son-of-a-bitch do that." (Dalrymple, 1989, quoting Sam Collier)

Amory Petroleum Company commenced a second well, the No. 1 Hall, one mile east of the discovery well, in December 1926. The well was plagued by mechanical difficulties. After two years, the hole was abandoned at 2,400 feet following a 13-month "fishing" job (i.e., recovering equipment lost in the hole). While the Hall well was being drilled, a third well was started in February 1927. Located one mile west of the Carter No. 1 discovery well, the Amory Petroleum Company drilled the No. 1 Dill well, resulting in a dry hole. The Benedum-Trees deal probably looked very good at this stage.

By then the trustee group had made a deal with a group from Chatham, Ontario, headed by R. L. Pattinson of Chatham and S. S. Reesman of Tulsa, Oklahoma. The Canadian group agreed to build a pipeline from the Carter well to the town of Amory and to buy gas from Amory Petroleum at "eight cents per thousand cubic feet," a fair price at the time.

A new company was formed, the Amory Natural Gas Company, with Pattinson as president. This was a separate company from the Amory Petroleum Company with entirely different ownership. The new company also obtained the franchise to supply gas to Aberdeen and Tupelo. In the first 18 months of sales, the Amory Natural Gas Company used 77 million cubic feet of gas from the Carter well, which it sold to domestic users, then totaling 600 in Amory (Steffey, 1929). At eight cents per MCF, this provided a total gross income of $6,160 for the 18-month period, hardly a windfall.

> After a few weeks when the expected riches hadn't materialized, Sam went by the blacksmith shop and Hoss-head said, "Sam, you damn fool, we shoulda' put up that hamburger stand when we had the chance."

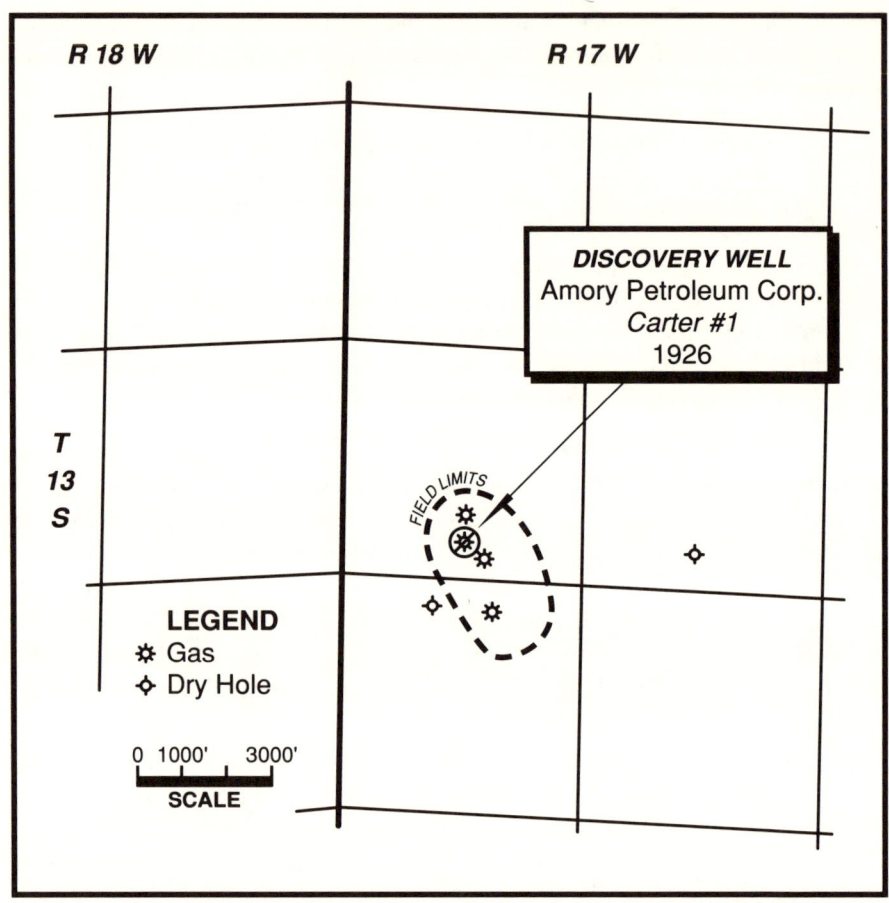

R 18 W   R 17 W

DISCOVERY WELL
Amory Petroleum Corp.
Carter #1
1926

T
13
S

FIELD LIMITS

LEGEND
✿ Gas
◇ Dry Hole

0 1000'   3000'
SCALE

FIG. 2   Amory Field, Monroe County, the first petroleum discovery in Mississippi
in 1926. All wells were depleted and abandoned by 1938.

Apparently, the Amory Natural Gas Company agreed to do additional drilling and began their Carter No. 1 some 600 feet north of the discovery well
on January 28, 1929. The well was completed as a gas producer on September 4 of that year at a depth of 2,582 feet. The initial production was only
149,000 cubic feet, and the "rock pressure" 600 pounds.

Commencing on August 9, 1929, the city of Amory was supplied by this
second well (owned by the Canadians), while the discovery well was shut

in. The well easily supplied the 83,000 cubic feet per day required by the town. At eight cents per thousand, this amounted to a gross income of only $200.00 per month, a paltry cash flow which continued for many months.

Since closing their deal with Amory Petroleum Company in 1926, the Canadian group had come under increasing pressure to supply gas to Aberdeen under the terms of the franchise. By 1930, antagonism was developing toward the Canadians from all sides. Several offers had been made by other companies to drill on the Amory Petroleum 36,000-acre block, but all the leases were committed to Amory Natural Gas, which refused the offers. The Canadians were sued by a law firm in Aberdeen to collect legal fees and by P. J. MacAlpine of Amory to collect drilling fees.

At the time, Southern Natural Gas Company was building a gas pipeline through northern Mississippi from Monroe, Louisiana, to carry gas to Atlanta. In a surprise move, Southern Natural made an offer in July 1930 to buy out the Canadians for an undisclosed sum (but guessed to be $400,000 as the Canadians were asking $500,000) and to lend Amory Petroleum Company $50,000 to drill more wells in the Amory Gas Field. The loan was to be repaid out of gas sold. The deal was closed on September 3, 1930. Thus, Southern Natural became owner of Amory Natural Gas Company and was granted the natural-gas franchises for Aberdeen and Tupelo. Time may prove the Canadians came out better than anyone in the deal.

Southern Natural Gas merged the Amory Natural Gas Company, into a new public utility firm, the Mississippi Public Service Company, with W. Rawson Collier as manager. This merger combined the gas distribution system of Columbus, Hattiesburg, Meridian, West Point, Macon, Starkville, and Louisville with the Amory system.

To accommodate the other franchises, a ten-inch gas line was constructed a distance of 30 miles from the Amory Gas Field to Tupelo with a four-inch spur to Okolona and a six-inch lateral line to Aberdeen. The city of Aberdeen began receiving gas on May 15, 1931, after a lengthy rate dispute. New drilling was started in the Amory Gas Field with expectations to eventually reach 200 million cubic feet per day and supply many of the towns in northern Mississippi. The declining pressure on the two existing wells was not considered relevant. The Carter No. 3 well was begun October 16, 1930, by Amory Petroleum Company and completed January 6, 1931. A small producer of 321,500 cubic feet per day, the well was located one-quarter mile east and slightly south of the discovery well. Next, the company drilled the Nell Harris well approximately one-half mile south of the discovery well.

It was completed on April 5, 1931, for 5,598,000 cubic feet of gas per day. This made a total of four gas wells in the field covering about 200 acres.

With the Amory Field now supplying Amory, Tupelo, and Aberdeen, the field pressure dropped very rapidly. By December 1933 the supply was not sufficient to service all of its customers, forcing the Carnation Milk Plant at Tupelo to be taken off-line and converted to coal. By 1934 all commercial and industrial users had been cut off and domestic consumers rationed. A compressor was installed to increase the flow of gas, but it was obvious that the field was being quickly depleted.

To supplement the diminishing gas supply, the Mississippi Public Service Company constructed a six-inch line from Aberdeen to tie into its West Point system, which was supplied by gas from Southern Natural. By December 12, 1934, the Amory system began receiving Monroe gas from Southern Natural, no doubt at a higher price.

A small amount of gas continued to be produced by the Amory discovery well until January 4, 1938, when it was taken off-line. All other wells had long since gone dead.

> Mr. H. G. D'Spain, General Manager of the Mississippi Public Service Company, Amory [in 1943], stated that the total production of the Amory Gas Field was probably not more than 1,500,000,000 cubic feet, including the 500,000,000 to 540,000,000 cubic feet estimated to have been wasted during the three months when the discovery well was allowed to run wild. (Vestal, 1943)
>
> Actually a total of 960,926 MCF of gas was sold from the field. (Dixie, December 14, 1944)

Steffey reports that the total money received from the sale of gas from the Amory Field to April 1936 only amounted to $59,000, at which time the field was almost depleted. The No. 1 Carter discovery well would have easily drained all the gas from the 200-acre reservoir.

The four producers and two dry holes probably cost over $100,000. These costs, along with operating expense, royalty payments, and ten years of overhead, resulted in a losing deal for the investors. The Amory Petroleum Company then took bankruptcy in early 1939, and all remaining assets were sold. Some thirty years later, Amory Field was to become a gas storage reservoir for Mississippi Valley Gas Company.

The sand member of Mississippian age from which the Amory Field produced was named the "Carter sand," and became the main producing formation for hundreds of wells to be drilled in the basin in later years.

Other Drilling in the Amory Area

After the discovery at Amory in 1926, several promoters rushed to the area and raised money to drill wells at random. All were dry, even after many months or even years of drilling. However, out-of-state oil companies invested in some of these ventures.

Clarence C. Day, a prominent lumberman and large landowner from Aberdeen, leased a large block to a Benedum-Trees subsidiary, the Arkansas Fuel Oil Corporation. In 1927–28 this company, with backing by the Texas Company and Transcontinental, drilled a 4,093-foot dry hole in Lowndes County near the junction of the Tombigbee and Buttahatchie rivers. This was one of the deepest tests in the state at the time.

It appeared that another discovery had been made when a well blew gas about a mile north of the Monroe-Lowndes County line. This well, the Natural Gas and Fuel Corporation, Frank L. Rye No. 1, blew in on November 13, 1927, at 2,693 feet, making 3,180,000 cubic feet of gas per day with some water. From an original pressure of 980 pounds per square inch, the gas flow declined rapidly and then gradually died out.

In November of the next year, P. J. MacAlpine was engaged by Arkansas Natural Gas Corporation, a Henry L. Doherty company that later became a part of Cities Service, to deepen the hole to 3,710 feet. Overseeing the operation was George W. Colvin of Arkansas Natural Gas. The well was finally abandoned on September 9, 1929.

After the Amory discovery, another large oil firm, the Ohio Oil Company, became interested in the Black Warrior Basin. Ohio leased several thousand acres around the Rye well at $1.25 to $1.50 per acre. Dealings were through C. C. Day and Jack Mobley and supervised by Ohio Oil's geologist from Jackson, Mississippi, J. Stuart Mosson. Drilling did not take place until several years later.

Soon most oil companies lost interest in the Amory area. The ancient Paleozoic rocks were hard to drill. Wells took months, even years, with a cable tool inching its way downward to potential pay sands. The technical capabilities of the rotary rig in the 1920s and 1930s were not sufficient to penetrate the hard rock at a practical rate. The prospects of hitting gas were also less exciting or potentially profitable than those of oil.

CHAPTER 9

The Oil People Arrive

The small promoters of the southeastern states were being joined by major oil companies in the late 1920s. Independents incorporated a new company for almost every venture, so that company names abounded, many of which were almost identical. The large oil companies proudly displayed their names on all the wells they drilled and on their company cars.

In the late 1920s oil companies began to establish offices with resident personnel in the southeastern states. Mississippi was the natural choice for such offices since it was the state nearest to the bustling oil cities of Shreveport, Louisiana, and El Dorado, Arkansas.

Steffey Scout Report

On October 6, 1928, a scouting service was established for the Mississippi and Alabama areas by Robert L. Steffey, operating from the Edwards Hotel in Jackson, Mississippi. Steffey was an experienced oil scout from Tulsa who recognized the growing oil interest in the southeastern states and was able to line up a group of subscribers for his triweekly scout report. He soon included Florida in his area of reporting. Steffey was known to have a weakness for drink and women, but these traits did not deter him from being a "damn good" oil scout in his time. During 1928 and 1929, Steffey moved his headquarters to the Park Hotel in Amory, Mississippi. He then moved his operation to the Lamar Hotel in Meridian, and, later in 1930, back to Jackson, an indication of the uncertainty as to where the oil business would concentrate in the southeastern states.

Steffey dutifully reported the oil and gas activities of the southeastern states from 1928 to 1941. In colorful language sprinkled with oilfield terms he wrote accurate, timely accounts of everything of interest to the oil community. He took special care to mention the names of almost everyone involved and the deals being made.

A large influx of oil company employees into Mississippi began in 1926. Most of the large companies transferred experienced people from oil-producing states. There was also a group of regular transient oil people visiting from other petroleum centers, mainly Shreveport, Dallas, and El Dorado.

Because no significant oil or gas production had been found companies initially picked different cities for their headquarters, most commonly Meridian, before 1930. The Steffey Report of August 9, 1929, listed 17 large oil firms that were qualified for business in the state of Mississippi:

Amerada Petroleum Corp.	Magnolia Petroleum Co.
Arkansas-Louisiana Pipe Line Co.	The Pure Oil Company
Atlanta Oil Producing	Shell Petroleum Corp.
Carter Oil Co.	Sinclair Refining Co.
Dixie Oil Co.	Standard Oil of Kentucky
Geophysical Research Corp.	Texas Co.
Gulf Refining Co.	Tidal Oil Co.
Lion Oil Refining Co.	Vacuum Oil Co.
Louisiana Oil Corp.	

Ohio Oil Company

Ohio Oil Company was one of the first to establish resident personnel in the southeastern states following the Amory discovery. The company's landman, J. R. (Jerry) Whalen, was stationed in Amory at the time of the flaring of the Carter well in 1926. He remained in Amory for years until he was transferred to Mobile, Alabama, in the mid 1930s.

Other representatives of Ohio who spent time in Mississippi and Alabama were vice president J. R. Kerr; G. R. Steinborger; geologist and magnetometer operator, R. G. Copeland; and C. L. Moody, a geologist from Shreveport.

Gulf Refining Company

Of the many companies that had become interested in the oil and gas possibilities of the southeastern states, Gulf was to be the most active during the 1926–38 period.

In 1926, Gulf Refining Company opened an office in Meridian, headed by geologist B. W. Blandpied; scout Edward F. Warren; paleontologist C. D. Fletcher; and, later, landman W. M. Downer. Ed Warren was replaced by Tommy McAter in December 1929. During 1929 Gulf recruited a group of young geologists to do surface mapping. Several of them later emerged among the more famous geologists in the southeastern states, including M. E. (Bud) Norman, C. L. (Buzz) Morgan, J. T. (Sparky) McGlothlin, and Henry N. Toler. Other geologists in the group were B. G. Martin, Floyd M. Ayres, James Tierney, James D. Aimer, Victor P. Grace, Gail Montgomery, Felix Richardson, Hasting Faulkner, and Buford Miller. Another Gulf geologist, C. M. (Army) Dorchester, would become well known in the Mississippi oil scene. Although not in the surface-mapping crews, Army had worked the Mississippi and Alabama areas for Gulf out of Shreveport since the early 1920s.

In April 1929, Gulf sent two aviators to Meridian, J. P. Vestor of Houston, Texas, and W. R. Henderson of Jackson, to do aerial photography. Sidney E. Mix, chief geologist for Gulf from Shreveport, along with his assistant, Roy T. Hazzard, often visited the office. In addition to the surface geologists, Gulf had a number of magnometer operators working in the southeastern states, including L. H. McCutchins, W. E. Kurz, L. Bell, and Jack Richards.

Gulf would be rewarded for this early geological work many years later with the discovery of several major fields. The Steffey Report estimated that at the height of its activity in 1929, Gulf was spending the astronomical sum of $300.00 per day to maintain this large staff.

Bernerd William Blanpied laid the ground work for much of Gulf's later activities in Mississippi and Alabama. Born in Kansas in 1897, he completed his geological education at Wichita, Kansas, and Oklahoma University, receiving his BS in geology in 1917. After serving in France during World War I, he became a petroleum geologist with Pure Oil Company in 1919. Gulf hired Blanpied in 1922. He was moved to Meridian in November 1924 and transferred to Tupelo a year later. Blanpied was in northern Mississippi at the time of the discovery of the Amory Gas Field and experienced the excitement that resulted.

In May 1929 he returned to Meridian to head up Gulf's surface-mapping crews. Three years later he was moved back to Shreveport, and became Gulf's chief geologist in 1948. He retired from Gulf in 1960 after 38 years with the company. His work in Mississippi influenced, to a large extent, the great success that Gulf was to have in the southeastern states.

One of the first young geologists recruited by Gulf for the purpose of surface mapping in Mississippi was Marvin Eugene (Bud) Norman. After graduating in geology from Texas Christian University in 1929, Bud was working on his master's degree when he was hired by Gulf.

Traveling by passenger train out of Fort Worth, he arrived at Gulf's head office in Shreveport. There he met with Roy Hazard, only to be given traveling money and put back on a train to Meridian. On arrival, he met B. W. Blandpied, head of Gulf's Meridian office.

The tall trees in Mississippi were a novelty to Norman. Bud recalls walking through patches of virgin pine forests, where the ancient trees formed a canopy that allowed no undergrowth, and the forest floor was a carpet of pine needles. Soon after he arrived, so did other young geologists, most right out of college. The group spent a few days in Tuscaloosa at the University of Alabama being indoctrinated on the surface formations in the area.

Bud recalls that on first arriving in Meridian, the geologists lived in the Lamar Hotel at $2.50 per night and had their meals at the famous Weidmann's Restaurant. However, this plush existence was not to last. Soon Blandpied moved them to a roominghouse where they set up their drafting tables and were furnished lodging at $1.00 per night. This was before the days of air conditioners; Bud recalled, "We sweated a lot over our maps." Meals were then furnished by a nearby boardinghouse which prepared sack lunches each morning for the men to carry to the field.

Two-man teams were assigned areas for surface mapping and commenced to work seven days per week. Bud started at Stafford Springs in Clarke County and during the ensuing months followed the outcrop of Yazoo Clay all the way to Philadelphia, Mississippi. His partner for the first few months was Felix Richardson and later Phil Montgomery. The pairs started out early every morning cross-country through woods and marshes and worked until dark. Bud decreed this was one of the most satisfying experiences of his life.

Following his assigned surface layer, Bud's team discovered a wide arc in the Yazoo Clay outcrop circling the town of Heidelberg. This later proved to be the result of a large buried structure that formed the trap for the Heidelberg Field, to be discovered 14 years later.

In 1944, at a geological presentation at Biloxi, Tom McGlothlin gave Bud Norman credit for finding Heidelberg Field. Actually, Gulf checked the area with geophysical work to confirm the structure. But certainly it was Bud Norman's surface work that brought it to the company's attention originally.

On December 31, 1930, the mapping crews were discontinued. The deepening economic depression after the stock market crash of October 29, 1929, was taking its toll. Gulf terminated most of the new geologists, but transferred Norman to western Texas where he was employed as a production engineer through the Depression years. He was to return to Mississippi in 1941.

Gulf closed its Meridian office and conducted its southeastern states operations out of Shreveport for the next ten years.

Other Companies

By 1929 other companies had stationed resident personnel in Mississippi and Alabama. These included Sun Oil Company, with Donald J. Monroe, geologist; Pure Oil Company, with Gene E. Alcott, scout, and two other unidentified employees; Louisiana Oil Refinery Company, with R. A. Moore, geologist; Texas Company, with B. E. Brewer, geologist and J. J. Mason, landman; and Carter Oil, with E. V. Whitwell, geologist. Shell Oil had W. W. Waring, a geologist in charge of Paleozoic areas, stationed permanently in Amory, while its landman and scout, Frank G. Hulse, was located in Jackson.

Standard Oil of Louisiana, as owner of the refinery at Baton Rouge, was very interested in Mississippi and often had its landman, Harry Creighton, in Jackson along with geologists, Sidney A. Packard and Harold K. Shearer, all from the Baton Rouge office. The Lion Oil Refining Company had offices in Jackson by 1929, with W. H. Reed in charge. In late 1929, Humble Oil & Refining Company stationed Paul J. Fly in Hattiesburg, "to familiarize himself with the area." This would remain Humble's Mississippi headquarters even after most of the other companies moved to Jackson in later years. There were also a large number of other companies interested in the southeastern states during the late 1920s operating from offices in other states, such as Amerada from Shreveport and Magnolia from Dallas.

CHAPTER 10

The First Seismograph in Mississippi

The Amerada Petroleum Company pioneered the development of reflection seismograph and by 1927 had devised a means of determining the depth of underground formations by measuring sound-wave travel time. In 1929, Amerada began to use this new technique around the Yazoo City area. The company operated out of Shreveport but stationed geologist J. W. Kisling in Yazoo City for a prolonged stay. The president of Amerada, Everette De-Golyer, made visits to Mississippi to inspect the operation. The company hired James P. Evans of Shreveport to assemble two large blocks of leases in Yazoo County, one immediately east of Yazoo City, and one several miles to the south around "Tinsley Switch."

By early 1930, Amerada was satisfied that a structure existed almost directly beneath Yazoo City and moved in the largest drilling rig ever operated in the state of Mississippi. The location for its Campbell No. 1 was two miles east of Yazoo City. Tippett Drilling Company of Shreveport was contracted to drill the well, using a 122-foot steel derrick, twin engines, two boilers requiring 20 barrels of fuel oil per day, five-inch drillpipe, and "ten and one-half pound per gallon weight" drilling mud. After just 78 days of drilling, the well reached a depth of 5200 feet to set a new depth record for Mississippi. But no oil or gas was found. In view of the failure of this well, the company decided not to drill the block south of Yazoo City. Had they done so, they might have discovered the giant Tinsley Field (Evans, 1988).

Despite the failure of this well to find oil, by 1931, reflection seismograph was being used by several companies in Mississippi. Reflection seismograph would become the primary exploration tool of oil companies. Gulf and United Gas kept seismograph crews working in Mississippi almost continuously for the next two decades. The process was soon being used in Alabama and Florida as well. By the end of the 1930s, a great number of "shootings" had been completed in all the southeastern states, setting the stage for big discoveries to be made in the 1940s.

Chapter 11

Early Oil Plays in Mississippi

As the 1920s came to a close, oil exploration in Mississippi was underway with mounting momentum. Four principal areas were active: the Amory area in Monroe County, the delta area in Sharkey and Issaquenna counties, the Topton area near Meridian in Lauderdale County, and the Jackson Anticline in Hinds County. Scattered shallow wells were being drilled by various small operators in other parts of Mississippi, but these received little attention.

The growing oil business in Mississippi was reflected in several bills passed by the Mississippi legislature requiring permits to drill, the filing of logs with the state geologist, and other regulations. But all of these bills were vetoed by Governor Theodore G. Bilbo.

The southeastern states initially gave little notice to the October 1929 stock-market crash. These states, with largely agrarian economies, placed little significance on the happenings so far away in New York financial centers. However, the waves from the stock-market crash would be strongly felt in the southeastern states within the next year.

The Topton Area

After the gas discovery in Amory, the next excitement was caused by a well drilled in the vicinity of Meridian, which appeared to have found oil. If oil was discovered at Meridian, it would extend production in the Gulf Coastal Plain another 200 miles eastward. The well, drilled in 1927, was the Lauderdale Oil and Gas Company No. 1 Gunn well, known as the Topton well.

Financed by Mississippi businessmen, the well was drilled to a depth of 3,487 feet eight miles northeast of Meridian. Oil shows were reported in a sand believed to be of Eutaw age.

Several small independent oil promoters rushed in to exploit the publicity generated by the well. Some of the people involved in these plays were F. X. Gowans, A. B. Amkis, Jr., Fred J. Hughes, J. A. Smith, Charles Esceu, Clint Vinson, and L. M. Shadboldt. George P. Atkins and Joseph A. Baker, who had promoted the Topton well, bought new leases on their 18,000-acre block, as earlier leases had expired. They set out to drill a second well by the same group of "capitalists from Jackson," who financed the first well.

Gulf became interested in the project, however, after a favorable report by its Paleontologist, C. D. Fletcher. Gulf Refining Company soon had as many as twelve magnetometer operators working the Topton and adjacent areas.

The company offered a deal to Atkins and Baker, who readily accepted. Gulf agreed to buy 1,920 acres of their leases around the well at $5.00 per acre with an option for the balance. The deal included a stipulation that a 3,250-foot well was to be drilled one mile from the Topton well. A location was selected on the J. T. Lackey farm approximately one and one-half miles northwest of the Gunn well. The rig for the No. 1. J. T. Lackey well was shipped from El Dorado, Arkansas, in two railcars. The well spudded on July 4, 1929.

The word was out and oil people flocked in. The first group included Fred M. Haase of Shell Petroleum, Shreveport; Miller and Shearer of Louisiana Oil Refining Company; Gene E. Alcott, scout for Pure Oil; J. R. Whalen, landman for Ohio Oil; and B. E. Brewer, geologist for the Texas Company. The Palmer Corporation of Shreveport had scout George L. Gilbert in Meridian along with a magnetometer operator. Soon other oil companies sent representatives to Meridian to watch the well. These included E. V. Whitwell, geologist with Carter Oil; Earl Fox, scout and landman for Vacuum Oil; J. J. Mason, landman for Texas Company; Allen G. Thurman of Shreveport; and several Gulf representatives.

Later arrivals were Donald J. Monroe of Sun Oil; Frank G. Hulse, landman for Shell; J. S. (Shorty) Fox, scout for Arkansas Natural Gas; R. W. Williams and M. M. Smith, independents from Shreveport; George P. Stevens, geologist, and Gene Morehead, landman, for Sims Oil; Paul J. Fly, geologist for Humble; W. W. Rusk of Producers and Refiners; Joe B. Hurley of El Dorado, Arkansas; W. M. Redditt, former landman with Texas Company; and T. G. Hibbler, lease broker from Shreveport. For the first time since the Amory discovery in 1926 sufficient interest was being generated in a southeastern state to attract representation from almost all of the major oil companies.

On August 1, 1929, Henry H. Clement, driller on the Lackey well, became ill with whooping cough as did all his family, so that drilling was suspended at 1,733 feet until he could "get himself and his family in shape" (Steffey, 1929).

Drilling resumed on August 7. After several delays due to weather, fishing jobs, and financing, an oil show was encountered just below 2,800 feet. As Steffey reports:

> Chronologically, here are the happenings and facts: After taking a core of fine sand from 2800 to 2805 feet, gas began to show up on the ditch Saturday from 2805 to 2808 feet. . . . Early Sunday morning as the core barrel was again run and the moment the pump commenced, gas bubbles showed up. The formation was at first pretty hard and then alternately softer streaks developed. Along with the gas was some oil drops of which bursted. Four and a half feet was actually cored.
>
> Pulling out, the core barrel was found empty but greasy; with the exception of three inches of a calcareous sand. It was decided to attempt another core of a foot Monday morning (only one shift is working at the well) and then run the tester. Monday morning's core of eighteen inches grayish white calcareous sand with the bottom red shale. NO SALT WATER HAS BEEN DISCOVERED. . . . Monday afternoon in running the tester, the tool failed to seat properly but in opening up, the fluid on the outside of the drill stem fell rapidly and at the same time gas came up through the drill pipe and on the outside. B. W. Blanpied, geologist, along with Edward F. Warren, Jr., (both Gulf Refining Co.) and myself smelt the gas it has a petroleum odor. The demonstration was over in a few minutes due to the rush of mud.
>
> Breaking out the first fourble, the fluid was found that high up the string. Right on top of the mud were found chunks of the dark brown sand.

Samples of the oil sand were obtained by several companies, all of which confirmed that it was porous with good oil shows, according to Steffey:

> Members of the Gulf Geological Staff state that the sand has excellent porosity double that of certain fields of Arkansas; gas encountered with the sand would substantiate the fact that the horizon contains live oil.

With this encouragement, the decision was made to run casing to test the shows. A common procedure in 1929 was to ream the hole to a point just above the oil shows to provide a "casing seat." Reaming was begun in early

November and an 11 1/4-inch hole was reamed to 2,808 feet. Casing of 8 1/4-inch diameter was successfully landed and cemented at 2,795 feet by December 15.

Cementing consisted of mixing cement on the surface by hand and pumping the mixture down the drillstem to the bottom of the hole. The heavy weight of the cement would cause it to equalize on the bottom inside and outside the casing. Several days were usually allowed for it to set up and harden. At that point, the cement inside the casing, "the plug," was drilled out and the oil show below the casing would be tested "open hole" by bailing the mud out of the casing.

This open-hole completion had many pitfalls. If the hole was drilled too far below the base of the casing it could expose water sands whose flow would be difficult to shut off. If the cement did not set up evenly around the outside of the pipe it might allow water sands behind the casing to flood into the well bore and fill the hole with saltwater.

With casing set, the second Topton well was poised to be the first oil discovery in the southeastern states. Expectations were arousing a great deal of activity including leasing and mineral buying. With this excitement, the well's operators, Atkins and Baker, took the opportunity to raise more capital and elected not to complete the well until after the Christmas holidays. Hopes were shattered on April 16, 1930, when the plug was drilled on the Lauderdale Oil & Gas Lackey No. 1, only to hit saltwater. Representatives from many companies witnessed the event. The well was drilled another 600 feet before the operator admitted defeat. The principal investor in the project, Gulf, then core-drilled the area and, finding no indication of structure, gave up its lease options. Interest in the area diminished and has stayed inactive ever since.

The Delta Play

In May 1928 a big gas discovery was made in East Carroll Parish, Louisiana, just 15 miles west of the Mississippi River. Palmer Corporation drilled the well, which tested 51 million cubic feet of gas plus a thousand barrels of saltwater per day, from a depth of 2,300 feet. Production came from the gas rock, the same formation that provides the reservoir for the huge Monroe Gas Field.

This discovery, known as the Epps Field, immediately set off a play in Mississippi in counties adjacent to the big well, primarily Sharkey and Issaquena counties north of Vicksburg. Located in the flat floodplain of the

Mississippi River, these counties were part of the rich delta farmland of Mississippi.

One of the first wells to be drilled on the Mississippi side of the river was by the O'Brien Brothers of Shreveport. Their Bellgrade Lumber Company No. 1 well, in Issaquena County, was spudded in May 1928 and completed on September 4 of that same year. A depth of 3,343 feet was reached with no oil or gas shows being encountered.

The Atlantic Oil Producing Company of Shreveport was also active in the delta in 1928. The landman, C. A. Strahan, leased 22,000 acres from George T. Houston & Brothers and the Frank B. Houston Estate in Sharkey County. Bonus paid was $26,000 with further consideration to begin drilling a well by December 1, 1928, or pay an additional $26,000.

It was necessary to lay a two-inch line from the railroad to fuel the rig. An eight-foot dam was also erected around the location to keep water out when the river flooded. The well was abandoned at 3,450 feet in the Selma chalk (according to Atlantic's paleontologist, S. H. Rook). The basis for locating the well was a magnetic high found by E. G. Nicar while working for J. P. Evans.

The Texas Company ran torsion-balance surveys in Sharkey County, in 1928 to check out two "highs" found by magnetometer. These were known as the Nitta Yuma and Richey prospects.

The Texas Company was trying to make a decision on an option held by the company from James P. Evans of Shreveport. Evans had leased a large block of acreage on both sides of the river known as the Lake Washington Block, and had granted the option to the Texas Company. For this option, the Texas Company paid Evans 50 cents per acre with an additional $1.50 to be paid for acreage selected by November 1, 1928. Records indicate that C. L. Goodson, landman for Evans, purchased royalty for $2.00 per acre. The Texas Company did not exercise the option, but United Gas drilled one of these structures in 1931 with no success.

By the early 1930s most companies were losing interest in the delta play, as it was overshadowed by more interesting events elsewhere. Even so, several more wells would be drilled in the delta during the decade.

CHAPTER 12

Natural Gas in the Southeastern States

The first important events in the petroleum industry of the southeastern states involved natural gas. The natural gas business was experiencing fast growth, capturing untapped markets in cities being supplied with manufactured gas. During the 1920s the first long-distance natural gas transmission pipelines were being built.

Many small gas companies had come into existence to serve local markets with limited sources of supply. The stage was set to combine many of the small systems into larger companies with broad-based supplies of natural gas.

Interstate Natural Gas

The first pipeline to transport gas from out of state into Mississippi was built in 1926. The Interstate Natural Gas Company constructed a 22-inch line that extended 170 miles from the Monroe Gas Field to Baton Rouge. It passed through southwestern Mississippi. The primary purpose of this line was to supply natural gas to the Standard Oil Company of Louisiana Refinery and to domestic customers in Baton Rouge and towns along the route of the pipeline.

> To investigate the Monroe gas field, as a suitable supply of low-cost fuel [for the Baton Rouge Refinery], they sought the advice of their parent, Jersey. And, Christy Payne, Sr., sent two of his top men to investigate. They were T. O. Sullivan, Vice-President and General Manager of the Hope Natural Gas Company, and Hope's Chief Engineer, Mr. Howell Cooper. (Davis, 1964)

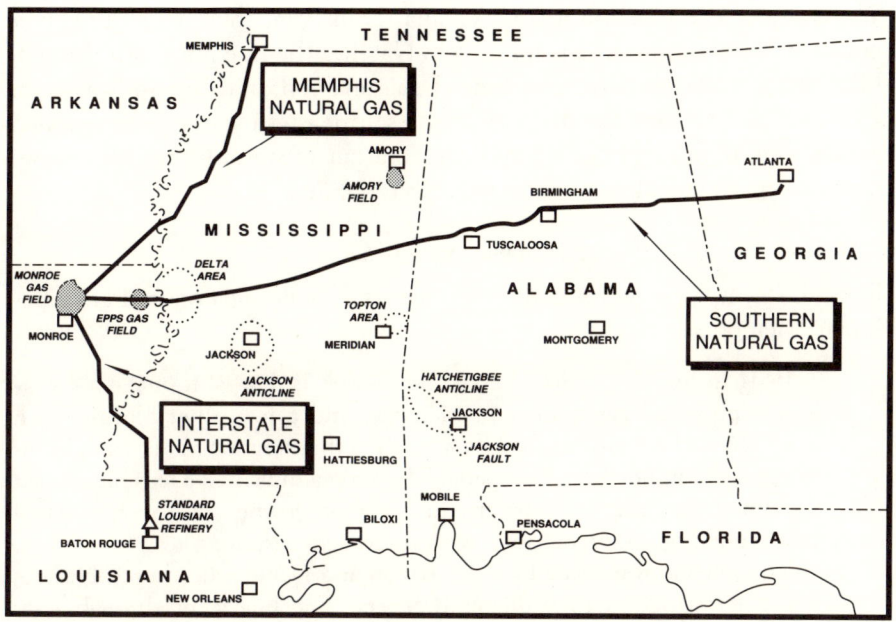

FIG. 3   Southeastern states, 1929; early natural gas pipelines and areas of most
intense interest for leasing and drilling

Payne then retained the services of Ralph E. Davis, one of the first engineers to develop methods to calculate accurately the reserves of natural gas
wells.

> We three men had a fine trip together. At the time there was a great
> flood in Louisiana, and the Ouachita River, which cuts through the gas
> field, and is normally about an eighth of a mile wide, was out of its banks
> and flooding the flat valley over many miles on each bank of the river. To
> get to individual wells, we went by motor boat and each well was tested
> as to its closed pressure and flow characteristics. We had a lot of informa-
> tion upon which to base a favorable report. The use of gas for carbon black
> in large volumes, and with modest decline in pressure, was the best evi-
> dence. I could see only a long life to a pipeline to Baton Rouge. . . .
> It was considered desirable to consider taking gas further south to New
> Orleans. We stayed in New Orleans long enough to get a taste of the
> quaint old city and to find out that there was almost no market there for
> natural gas. The winters are warm, and as for fuel-consuming industries,
> they had practically none. (Davis, 1964)

The route of the pipeline was to Vidalia, Louisiana, then across the river bridge at Natchez into Mississippi. From Natchez the line ran south through Mississippi, until it again crossed into Louisiana and on to Baton Rouge. In 1927 Natchez became the first city in the southeastern states to be supplied with a long-term supply of natural gas from out of state. Other Mississippi towns along the pipeline's course were also supplied.

Memphis Natural Gas

Ralph Davis was also consulted on the merits of building a natural-gas pipeline to Memphis:

> In 1926 he [H. G. Scott] asked me [Ralph E. Davis] to make a survey of the potential gas market in Memphis and a few other places which would likely be easily reached by a pipeline that might be built from Monroe gas field in northern Louisiana. The survey indicated a feasible project providing that the gas company in Memphis [owned by the Mississippi Power and Light Corporation] would become a customer. . . .
>
> The low bid was made by Ford, Bacon and Davis, who then sublet the actual construction to Williams Brothers. The line went through sixty miles of swamp, it crossed the Mississippi at Greenville, and then north to Memphis through the flat Mississippi farming country. . . .
>
> I put Lyon Terry in charge of the project, with headquarters at Greenville, and Faber Hanst in charge of compressor station construction. We had a fine team, and six months after I had been told to "go," gas was turned on in Memphis. The mayor of the city opened the valve on New Year's Day, 1928. It was during that experience that I [Ralph E. Davis] got well acquainted with the Williams Brothers, Miller and Dave. . . .
>
> The 218-mile [18-inch] line from Monroe Field to Memphis, through great swamps and across the Mississippi, was built in record time. . . .
>
> It put Williams Brothers into a position to continue in business and they became the most outstanding firm of pipeline constructors in the world. (Davis, 1964)

Gas was thus made available to Greenville and other towns along the route. When the town of Cleveland, Mississippi, was piped for gas from this pipeline in the hot summertime of 1930 or 1931,

> All of the ditches were dug by hand. A gang of men stood across the street waiting for someone to drop out to take over his job. (Kelly, 1990)
>
> We paid help 7 cents per foot to dig the ditches. A man was assigned a portion to dig. At the end of the day we would measure to see how many feet he had dug to determine what he was to be paid. (Lambdin, 1990)

United Gas

Second only to Gulf, the company that was to play a leading part in oil exploration in the Southeastern States during the 1930s was United Gas. This company was one of the first large natural-gas-transmission companies to be formed. It began as the Palmer Corporation of Shreveport with principals of the famous Palmer House Hotel in Chicago, Homore Palmer and Potter Palmer, as stockholders. The company was active in Louisiana before 1915 with gas wells and a short gas pipeline to Shreveport. At the age of 22, Noriss Cochran McGowen, known as N. C., was sent to Shreveport as manager. After the discovery of the Monroe Gas Field in 1916 the Palmer Corporation acquired control of a large number of gas properties in the field. In 1928 the gas interests of the Palmer Corporation and those of Electric Power and Light were combined to form the Louisiana Gas and Fuel Company. Electric Power and Light Company had been operator of the gas-fueled power plant in the Monroe Field that supplied Jackson with electricity.

Even before this merger, the Palmer Corporation had scout G. L. Gilbert traveling through Mississippi and Alabama as early as 1928. By 1929, under the auspices of Louisiana Gas and Fuel Company, an office was opened in Meridian, with Gail F. Moulton, geologist, in charge and Jackson S. Young as scout. Both answered to John Ivy, chief geologist in Shreveport.

The Electric Bond and Share Company of Chicago was the major owner in the Louisiana Gas & Fuel Company, and N. C. McGowen moved up as head of this larger company. He soon combined the assets of several other gas systems including:

- Ouachita Natural Gas Company, with a pipeline from Baton Rouge to New Orleans,
- The Magnolia Petroleum Company gas interest (already separated from the parent company into Magnolia Gas Company) with pipelines from Shreveport to Dallas, Fort Worth and Beaumont,
- The Moody Seagraves interest, with gas wells in South Texas and pipelines from Corpus Christi to Houston, Houston to Shreveport, San Antonio, and Austin,
- North Texas Utility Company, with a pipeline from the Texas Panhandle to Wichita Falls, Texas, and having a number of its own natural-gas wells.

These companies, along with others, were officially combined into the United Gas Corporation on June 30, 1930.

It took several years to complete these mergers, but by late 1937 the many entities under the United Gas umbrella were combined into only three companies: United Gas Pipeline Company, which controlled the long-distance-transmission pipelines; United Gas Corporation, the distributing company for cities; and Union Producing Company, the oil and gas exploration and production company (Stotz, 1938).

Early wells were usually drilled under the auspices of United Production Corporation, United Gas Public Service, or United Gas Systems. United Gas was very active in Mississippi during the development and marketing of the Jackson Gas Field.

Southern Natural Gas

An entity conceived in the southeastern states and destined to become a major natural gas company was Southern Natural Gas Corporation. In 1928 Adam H. Davidson envisioned the building of a natural-gas pipeline from the Monroe Gas Field of northern Louisiana across Mississippi and Alabama to Atlanta. This giant project involved building a 460-mile pipeline, quite an undertaking at that time.

Financing was arranged by G. L. Ohrstrom & Company, and the Southern Natural Gas Corporation was incorporated in Delaware on May 9, 1928. Arthur L. Mullergren reported on the feasibility of the project.

> The market surveys were carried out by a group of five men, who began work in March, 1928. These men were George C. Maples, J. B. Milmoe, Louis Lyons, Jr., F. W. Gaw and Ernest Moeller. . . .
>
> Little of the present wealth of technical information on gas combustion was available, each problem being solved in front of the boiler of kiln with hours of trial and error. To make many of the conversions to natural gas Moeller devised his own burner, which was most successful.
>
> The original survey, made by these men, consisted of traversing the general territory by car, keeping a watchful lookout for signs of black smoke belching from industrial stacks. The blacker the smoke, the less efficient the combustion, and consequently the more savings that could be attributed to natural gas. . . .
>
> Construction was begun on May 1, 1929. The first 130 miles east from Monroe was completed by September 24, 1929. . . . Gas was first delivered into the Birmingham area on December 31, 1929. . . . It was estimated that as many as six thousand men were employed in the construction of the pipeline. An article in the Jackson (Miss.) Daily News. . . . stated, "The Corporation is paying fancy wages. . . . the average being about $4.00 per day."

A gala banquet to celebrate the completion of the line was held in Birmingham on January 29, 1930. . . .

A sleet storm struck Birmingham, and the streets and roads over the mountain were glazed with ice. The banquet had been scheduled to be held at the Hollywood Country Club in Shades Valley and all of the food was prepared. All of these distinguished guests would have to drive over these dangerous hills and ice covered roads to get to the club. This was too great a risk. The Fire Department was called and asked to clear the streets, and with great animation and good will went to Twentieth Street hill and turned on their fire hoses. The more the water was applied to the icy street the more icy they became and the more pressing the problem.

Southern Natural was a new company, but even then it could show resource and imagination. Out came the red [Southern Natural] trucks, complete with chains, and over the mountain they went and returned, bringing back to the Tutwiler [Hotel] the fruit cocktail, the filet of trout, the broiled chicken, the frozen delight, the roquefort cheese and the cigars and mints. The day was saved. The dinner went on. (Ivey, 1967)

As 1929 ended an army of pipeliners was laying pipe at breakneck speed toward Atlanta. However, the 1929 market crash had begun to make financing difficult. Construction bills were piling up, an omen of disaster.

In addition to the main line and many short lateral lines, Southern Natural simultaneously constructed a long trunk line from Benton, Mississippi, to Mobile, Alabama, and in 1930 extended it to Pensacola, Florida. A second major trunk line was constructed from Tuscaloosa to Montgomery and on to Columbus, Georgia. With the Depression becoming more severe, the company was unable to raise additional money in the public market.

By March 31, 1931, construction costs had amounted to $38,885,983.75, which was some $4 million more than the cost projected by Mullergren. In order to stave off bankruptcy, some of the assets were sold. Nevertheless, the company was forced into receivership on October 1, 1931, but managed to restructure and emerge on November 1, 1935, with a slight name change to Southern Natural Gas Company (rather than Southern Natural Gas Corporation). No drilling or exploration was attempted by the company until 1936.

CHAPTER 13

The Jackson Gas Field

A geologic event that occurred some 65 million years ago was to set the stage for the first significant petroleum discoveries in Mississippi. All of Mississippi then lay under water except the extreme northeast corner of the state. In this sea, where the city of Jackson is now located, a volcano erupted and formed an island several miles across. As centuries ticked by like seconds and the volcano became dormant, the island began eroding and a halo-shaped atoll reef developed around it. Eventually the fast-growing reef closed in on the retreating shoreline to cover the island as erosion carried it below sea level. The reef fanned out thirty miles to the northwest, with the edge reaching as far as present-day Yazoo City. Simultaneously, a similar reef was developing a hundred miles to the west. It would become the Monroe Gas Field.

As millions of years passed, the seas advanced and retreated time and time again, leaving layer after layer of sediments over the old reef until it was buried to depths of 2,300 feet at Jackson, dipping to almost 5,000 feet near Yazoo City.

When humans first appeared on the scene, the present landform had developed, and the Pearl River wound its way through the state to the gulf. Compaction of sediments over the old buried volcanic island and reef created an area of high ground through which the river flowed. The high banks made a good boat landing and campground, known as Lefluers Landing. Over the years, the city of Jackson spread out from there.

The presence of the Jackson Uplift was first suggested in a publication by

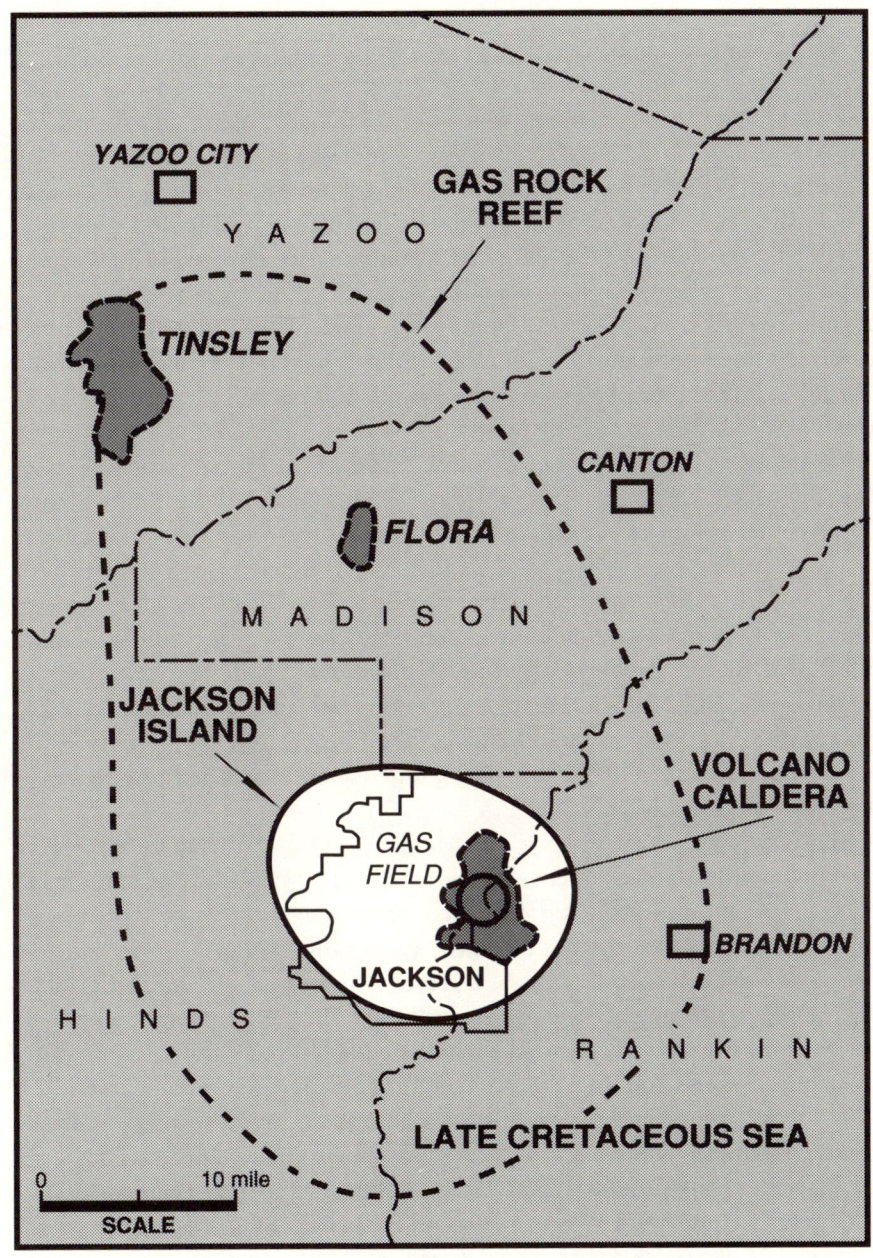

FIG. 4 A volcanic island which formed in an ancient sea some 65 million years ago is buried under the city of Jackson, Mississippi. An atoll reef which developed around the island provided the reservoir bed for the Jackson Gas Field and the Tinsley and Flora oil fields.

E. W. Hilgard in 1860, entitled "The Geology and Agriculture of Mississippi." The U.S. Geological Survey made a study of the Jackson area in 1915 and confirmed the presence of the anticline. Its 1916 publication, U.S.G.S. Bulletin 641 by O. B. Hopkins, suggested that the structure might have oil or gas possibilities. This publication called the attention of the oil companies to the huge structure and soon two companies moved in and drilled.

The first of these was the Atlas Oil Company. Its Garber No. 1 well was drilled to 3,079 feet at a location just northwest of Jackson and abandoned in June 1917. Benedum-Trees, famous independents from Pittsburgh, made a second attempt just north of Jackson, which would later prove to be within one mile of the edge of the Jackson Gas Field. The well was abandoned at 3,043 feet in October 1917.

No other drilling is known to have taken place on the structure until 1925, except a very shallow well drilled to 1,413 feet on the Elton Plantation six miles south of Jackson in 1920.

Ella Rawls Reader Stokely

The next drilling on the Jackson Dome was done by an incredible woman, Ella Rawls Reader Stokely, who had achieved worldwide recognition as an entrepreneur and businesswoman around the turn of the century. In 1905, *Everybody's Magazine* declared Mrs. Reader "the greatest businesswoman in the world." The story of her accomplishments was published in a four-part series in the monthly publication.

Ella Rawls was born in 1873, at Morgan Springs, Alabama, moving to Birmingham in her early teens. In 1891, at 18 and penniless, she moved to New York City, but soon started a stenographic agency, which within four years had grown to be the largest in the metropolitan area.

She devised a unique technique to record various City Commission hearings, court sessions, and political meetings. Using relays of stenographers, made up mostly of men, she had each stenographer record some 15 minutes of the proceedings and then stop to type his portion with carbon copies. The copies were immediately passed on to a battery of typists to make multiple copies. With this procedure Ella Rawls was able to have numerous copies of a transcript of the proceedings completed almost as soon as the hearing was concluded. These were in great demand by newspapers. Her steno service was such a huge success that she had made a fortune by 1900.

In 1901 she married Athole B. Reader, a New Zealander of English parentage and an international businessman. Mrs. Reader turned to other entrepreneurial projects, including organizing a $10 million railroad project, obtaining various rail-related contracts and concessions in London and India, and organizing a copper conglomerate in Peru. She became politically involved in South America and was soon adviser to the head of Santo Domingo, to the chagrin of President Theodore Roosevelt who felt Mrs. Reader was meddling in foreign affairs. Nevertheless many members of Congress listened to her advice. She successfully pursued her mining interests into Canada and organized a large powdered-milk plant in Camden, New Jersey.

After she was twice widowed, (her second husband was W. L. Stokely), she moved back to New York. Through research, she became convinced that Mississippi would be the next oil state and the Jackson Dome was the most prominent structure in the state. Following her instincts, she came to Jackson and prepared to search for oil. The fact that she had never been involved in drilling for oil was of no concern to her. Reader also brought her niece Frances Stewart and Frances's young daughter to Mississippi (Pepper, 1989).

On December 5, 1924, Ella Reader obtained a lease on the 1,300-acre State Insane Asylum lands (now the University Hospital-R&D Center Complex) for $1.00, with the obligation to commence a well by March 5, 1925. The lease was granted by the asylum's Board of Trustees.

Reader began drilling in March 1925 off Old Canton Road near the present site of the Smith-Wills Baseball Stadium and Agricultural Museum. The well reached a depth of 1,700 feet in late 1928, when the derrick was lost to a windstorm and had to be rebuilt.

The validity of Reader's lease was challenged in court under the allegation that the Board of Trustees of the Insane Asylum Hospital did not have the authority to grant a lease on state lands. Reader lost in the lower court and on February 13, 1928, in the Mississippi Supreme Court. She elected to continue to drill anyway.

In the spring of 1928, she drilled through the gas zone that was to be discovered two years later by Cleve Love. "Gas showed up, but was mudded off," according to Steffey. Reader continued to drill with many fishing jobs and setbacks until the well finally reached a depth of some 2,900 feet by April 27, 1931, at which depth she abandoned the well. Altogether she spent some $40,000 on the project.

She continued to claim the lease and asked the district federal court for an injunction to restrain the state from leasing the land to other parties. The land was tied up by litigation until near the end of 1934, when she lost in federal court.

At 61, Ella Rawls Reader Stokely died at Jackson in November 1934. Steffey reported, "By all that is holy her well . . . should have been and could have been in proper hands the discovery well of the Jackson field. . . . Her well is now surrounded by producers." Her niece, now Frances Pepper, became a permanent resident of Jackson, as did Frances's daughter, now Frances Michel.

The Misterfeld Well

During drilling on the Reader well, other oil company teams began to show an interest. Gulf had two magnetometer operators working the Jackson structure in November 1928. Sun Oil had Donald J. Monroe doing surface work while the Palmer Corporation kept G. L. Gilbert scouting the area.

The next attempt to find oil or gas on the Jackson Dome was made by Lion Oil Company. Lion took over a block of 16,000 acres from Cleve Love and R. A. Reaton in Rankin County, some five miles due south of the city. Lion sold "units" in the prospect to major investors, mostly other oil companies. The No. 1 Misterfeld well was commenced in March 1929. While the well was being drilled, activity picked up around Jackson.

Gulf increased its number of magnetometer operators to six in addition to two torsion-balance crews working the structures. W. W. Downer with Gulf, Jack Hesterly with Louisiana Oil Refining Corporation, and others were buying leases. Mike McKay and Lon Mason bought a spread of minerals for Magnolia Petroleum Company.

The Misterfeld well ran "500 feet low" to the Reader well and was plugged in June 1929 at a depth of 4,075 feet.

The Rainey Well

Not discouraged, the Love brothers, headed by Cleve Love, and their major investor, Stewart Gammill, formed the Home Oil Producing and Refining Company, with a capitalization of $100,000, and began buying leases on the structure. The basis for locating the next well was a strong magnetometer high at Flowood, Rankin County, just south of Highway 80 (now old Brandon Road). This fell one and a half miles east of the Pearl River railroad bridge at Jackson.

Stewart Gammill of Jackson put up $10,000 and other businessmen another $5,000. Lion Oil contributed $10,000 plus a thousand acres of leases to the venture. Gulf took a $4,000 interest in the project and used its magnetometer information to help select the location. Gulf sent Army Dorchester to "sit on the well." Pure Oil gave $1,000 and Lion Oil $1,500 in dry-hole money, i.e., to be paid only if the hole was dry. Atlantic and Carter were also involved.

The Home Oil-Rainey No. 1 well was spudded in mid-September 1929, with the Love brothers as operators. On October 11, 1929, the well encountered the gas rock at 2,509 feet, some 578 feet structurally higher than the Lion Oil Misterfeld well. Gas showed up in the mud, but the well "lost returns" and all the mud in the reserve pit was pumped down the hole with none returning to the surface. Sawdust and hay were added to mud and pumped into the hole in an attempt to seal off the leaking crevice or fissure. One hundred sacks of cement were also pumped into the hole. While this was taking place, 8 1/2-inch casing was stacked on the location to run in the well if mud circulation could be resumed.

The showing of gas in the Rainey well caused considerable leasing and royalty buying in Rankin County and throughout the Jackson structure. Prices were reported as high as $12.50 to $15.00 per acre for royalty. Present on the well were Mix, Dorchester, Warren and Blocker for Gulf; Stathers and Steffins for Standard Oil; Fox for Vacuum Oil; Kisling for Amerada; Mason for the Texas Co.; Monroe for Sun Oil; Hulse for Shell; Whitwell with Carter Oil; Stevens, scout for Quachita; Wood, landman for Magnolia; Moore for Louisiana Oil and Refining; Paul Fly with Humble; and Gilbert, scout for the Palmer Corporation. None of the companies wanted to be left out if a major find was about to take place.

After many days, the lost circulation problem on the Rainey well became progressively worse, in spite of valiant attempts to correct it. The hole was abandoned on October 29, 1929, with the intention of skidding the rig and drilling a new hole. By coincidence, this was the date of the stock-market crash in New York. The loss of the well did little to diminish the enthusiasm of the oil companies and speculators. However, it was several months before a second attempt was made on the Rainey lease.

The Mayes Well

Undaunted by the Rainey well's troubles, Cleve Love and his associates began preparations to start a second well some two miles to the north, across the Pearl River in Hinds County. This well was to be on L. L. Mayes's land in the river bottom approximately one mile behind the old State Capitol Building (immediately across I-55 from the Downtown YMCA today).

A new company was to be formed to handle the project, with 30 units sold at $500.00 each. Mayes and Sidney McLaurin were associated with this new venture. Timber was moved in to start the derrick in early November 1929, but the Pearl River flooded, forcing them to move to higher ground. This was before the levee was built.

In the meantime, Gulf was negotiating to take over Home Oil's Rainey block. Gammill, who had been Cleve Love's chief backer and president of Home Oil, was very disillusioned at the outcome of the Rainey well. He released the rig to the Love brothers and sold the pipe and other equipment stacked on location to the Lackey well project at Meridian, where he also had a large interest.

Gulf proceeded to take control of the Rankin County portion of the Jackson structure. In late November 1929 it bought a large portion of Lion's holdings from the manager, John H. Ganzel. Gulf negotiated a deal with Home Oil to take over its 5,000-acre block around the Rainey well for a 25 percent net profit interest, and Gulf added another 7,000 acres it acquired on its own. Gulf's obligation was to drill three wells no more than 90 days apart. Gulf planned to skid the old wooden derrick, but then decided to move in a new rig capable of drilling to 6,000 feet.

The race was on to see who would establish the first production on the Jackson structure, and, it was hoped thereby be awarded the natural-gas franchise to the city of Jackson. Cleve Love appeared to take the lead when he spudded his L. L. Mayes No. 1 well on Wednesday, December 4, 1929, at 4:45 P.M. By December 30, 1929, the well was drilling at 1,700 feet, while Gulf still had not completed negotiations on starting the second Rainey well.

Oil people continued to pour into the Jackson area. Petro Royalty Company of Tulsa had its geologist, Willis Storm, in Jackson and its landman, Harper, buying mineral rights around the city. Thus, 1929 ended with Jackson engulfed in an oil and gas play of frenzied proportions.

On New Year's Day, the Mayes No. 1 well twisted off while drilling at 1,768 feet, causing a day's delay. Nevertheless, the well reached the gas rock at 2,461 feet on January 12. The operator proceeded to run and cement 6 5/8-inch casing to total depth. While waiting for the cement to set, Love took the opportunity to bring in more investors before the plug was drilled. Steffey reports, however, that he "paid his help in full to date."

Love proceeded to incorporate the Jackson Oil & Gas Company, a Louisiana corporation licensed to do business in Mississippi. This is not to be confused with C. R. Ridgway's old company, which had been liquidated.

The principal investor in the new corporation was W. C. Feazel of Monroe, Louisiana, who was made president. Other investors were W. T. Murray, vice president; M. L. McCully, secretary and treasurer; and Cleve Love and C. E. Malley, directors. After casing had been set, the well was running 70 feet "high" to the Home Oil-Rainey well. This was an opportune time for Love to bring the new capital into the deal.

Plans were made to resume drilling February 3, but a lack of coal delayed them. One of the principals was also away on a hunting trip and wanted to be on hand when the plug was drilled. On the night of February 12, however, steam was up and drilling was commenced on the plug. By February 16, the hole had been drilled to a depth of 2,585 feet, with a show of gas beginning at 2,525 feet.

Realizing that the "pay zone" might have been drilled unintentionally, the crew began bailing the mud from the well until it suddenly blew out at three o'clock Sunday morning, February 16, making some 2 million to 6 million cubic feet of gas per day with considerable saltwater. Rock pressure taken by W. H. Vaughan of Gulf Refining was found to be 820 pounds per square inch.

E. O. Spencer had taken the plush Panama Limited to New Orleans on Sunday morning only to learn of the discovery from the *Times Picayune* at the New Orleans train terminal:

> Local people were excited, but were now being cautioned by the Jackson Daily News "To Keep Your Head." . . . Mineral rights near the Mayes is reported priced at $25 per acre. (Steffey Report)

As usual after such an event, many oil people rushed to Jackson. Already there watching the well were B. E. Brewer (Texas Company), Jackson Young (Louisiana Gas & Fuel Corporation), Don Monroe (Sun), Army Dorchester (Gulf), Paul Fly (Humble), and John H. Ganzel (an independent). Johnny Rogers, landman for Louisiana Gas & Fuel, rushed over from Shreveport to buy leases for his company.

The Second Rainey Well

Gulf Refining Company now hurried to begin its well on the Rainey lease at a spot only 260 feet west of the Home Oil Producing Corporation's No. 1 Rainey. In early February, Gulf shipped a steel derrick to Jackson along with other heavy drilling equipment. The wooden derrick of the Home Oil well was dismantled to obtain lumber for the construction of a loading platform and coal chute on the railroad spur near the new location. The well was spudded at 4:00 P.M., on February 19, 1930, and reached a depth of 2,520 feet by March 16. A drillstem test on the gas rock at 2,500 feet to 2,520 feet recovered drilling mud, gas, and saltwater cut with 40 percent crude oil.

The showing of oil in this well brought an influx of scouts, geologists, and landmen who began leasing up every tract available in and around Jackson. Oil was much more profitable than gas. Prices ranged from $25.00 to $40.00 per lease acre. J. P. Evans took 8,000 acres west of Jackson around Bolton. However, after nine-inch casing was set and the plug drilled, the well flowed an estimated 15 million cubic feet of gas per day with some 10,000 barrels of saltwater with only one percent heavy tarlike oil. Efforts to shut off the saltwater were unsuccessful and the well was deepened. After no deeper producing zones were found, the well was plugged back to the gas rock zone. Following many more weeks of cementing and testing, the well was finally abandoned.

Jackson's Changing Skyline

The boom in Jackson began to pick up. New locations were staked daily. By April 10, 1930, 14 derricks dotted the skyline. In early March, Louisiana Gas & Fuel Company, through its agent and attorney, Garner W. Green, had made a location one-half mile north of the Mayes discovery well on 35 acres belonging to Mrs. Sollie McWillie Harris. This well was dry.

A brother of Cleve Love, Trigger Love, incorporated the Love Oil and Gas Company with a capitalization of $50,000 to drill a well on a 13-acre tract half a mile north of the Mayes well. His well was subsequently dry and abandoned.

The Jackson Oil & Gas Company flush with success after the Mayes well discovery, quickly located a second well approximately half a mile to the north, on the Swep Taylor farm, in February 1930. L. L. Mayes, landowner under the discovery well, donated gas from his well for the drilling of the Taylor well, and a gas line was laid to this new site. However, the rig that drilled the discovery well had been reserved by "Big Boy" Love, another of Cleve's brothers, who chose to use it on a location of his own. A distant rig had to be moved in from Louisiana for drilling the Taylor well, which also resulted in a dry hole. Altogether, the Jackson Oil & Gas Company would eventually drill a total of eight wells, all of which would be dry except the discovery well and the Mayes No. 4, which was drilled in 1932. Bad luck plagued the company.

The momentum of the play seemed to make people oblivious to the fact that only one marginal gas well, the No. 1 Mayes, had been completed from the first five wells drilled. Six dry holes had also been drilled, outlining the structure before the Mayes well.

The Mighty Mendoza Well

The next well drilled would prove that a significant gas field was present under the city of Jackson. Big Boy (L. P.) Love formed the Love Petroleum Company with his brother J. W. "Peanut" Love, E. O. Spencer, J. A. Roell, Garner Green, and others. The company began to drill its No. 1 Mendoza well approximately one mile east of the old Capitol building in the Pearl River bottom. Today it would fall immediately east of the Pascagoula Street and I-55 intersection. On April 29, 1930, the well blew in at a rate of 28,569,920 cubic feet of gas per day, flowing with 1,100 pounds pressure, with no saltwater. This well would ultimately produce and sell more gas than any other well in the Jackson Gas Field.

Big Boy Love went on to drill some 19 producing gas wells at Jackson, and his company, Love Petroleum Company, became the operator of the largest number of wells in the field.

The Love Brothers

The Love brothers, four in all, arrived in Mississippi in the late 1920s. Cleve Love was the oldest, followed by J. W. (Peanut) Love, Wesley (Trigger) Love, and L. P. (Big Boy) Love. All were born in Arkansas, but moved to Blanchard, Louisiana, at an early age with their parents and sisters. The brothers first began working in the oil fields at Caddo, just north of Shreveport. Other oil plays then led them to East Texas and, later, Arkansas. From the oil boom towns of El Dorado and Smackover, Arkansas, the four came to Mississippi and began promoting well deals around Jackson. This activity resulted in the discovery of the Jackson Gas Field in 1930.

Cleve Love, a flamboyant wildcatter by nature, went on to drill many wells in Mississippi, Louisiana, and Arkansas after the Jackson success. He is said to have made three fortunes only to lose each. His last years were spent earning a moderate living by pumping oil wells in north Louisiana. To the very end, he never lost his fire and enthusiasm for the oil business. His tales of the early days were spellbinding.

Love Petroleum Company, under Big Boy Love's leadership, grew into a very prosperous oil company that was active in Mississippi for the following

35 years. *The Jackson Daily News* in its May 31, 1944, edition called Big Boy Love "the Father of the Jackson Gas Field." Another article in the same edition indicated that his Company was the only local oil and gas company to survive of all those formed during the Jackson Gas Field boom.

Peanut Love remained with Love Petroleum Company as president until 1937, when he sold his interest in the company and went into the real estate business. Trigger Love dropped out of the oil business after drilling his dry hole in the Jackson Gas Field following the Mayes discovery.

The operator with the greatest number of gas wells, second only to Love Petroleum Company, was Alexander Trust of Louisiana with 14 wells.

Stewart Gammill

Not to be left behind, Stewart Gammill, who had been involved in the Rainey well at Jackson and the Lackey well in the Topton area, now formed the Pearl Valley Oil & Gas Company. Pearl Valley acquired a thousand acres along the Pearl River between the Mayes well and Rainey wells.

A major participant in his new company was Johnny Glassell, a Louisiana operator, who moved in his own rig and commenced drilling the company's first well, the No. 1 Littlefield. On July 13, 1930, the well came in for 32 million cubic feet of gas per day. Subsequently, the company would drill and operate five gas wells in the field.

Stewart Gammill, having many irons in the fire, also had under construction the Robert E. Lee Hotel, at the time, so his cash outlays must have been substantial.

Governor Theodore Bilbo

With the Jackson Gas development leading the headlines, Governor Bilbo used his famous oratorical skills to appeal to the state legislature for an appropriation of $250,000 to drill gas wells on state land around Jackson. He did not count on the tenacity of Ella Rawls Reader in holding on to the insane asylum lease. In an address on February 24, 1930, to the legislature, the governor stated:

> It will be a great and glorious day for Mississippi and her taxpayers when the great gas and oil field is brought in near the capitol of our state. The taxpayers of Mississippi own in their own right and name between four and five thousand acres in the heart of the section where this long looked for field is to be developed.
>
> I have been slow in becoming excited over the many reports of the prospects of discovery of oil and gas in Mississippi territory, but the recent

test made in the gas well near the capitol, together with the judgment and conclusion of many experts, have almost persuaded me to believe in the certainty that the Jackson territory is destined to become a great gas and oil center. For over two years I have persistently refused to give my consent as Governor of the state to many propositions to lease the state land on the usual one-eighth royalty basis for gas and oil. The oil experts and "wild-cat" drillers have seemed so certain that gas and oil were here to be found in great abundance, especially beneath the state's property that I have felt it would be almost criminal on my part to sign away the rights of the people to speculators when such rights are destined to become worth untold millions to the state.

I have handled these properties for the state, not as I would have handled my own property, being a poor man, but I have handled them as I would handle my own property if I were a rich man, and thus being able to develop my own gas and oil prospects and share in the rich returns should it come. Mississippi is rich and is just as able to develop this certain field in oil and gas as any oil company or corporation in the world. Then, why should I, representing the state, surrender seven-eighths of the values of these prospective millions, in order that others may become rich, and deny the taxpayers the relief that would come to them if this oil field is a certainty? . . .

If a real gas and oil field is found in this territory, it is no idle dream to prophecy that the state's share properly safe-guarded would soon pay the state's entire bonded indebtedness and even be great enough to defray all of the state's expenses and make our state tax free in so far as state obligations are concerned.

In view of such possibilities permit me to suggest the wisdom and advisability of the Legislature appropriating about $250,000 and placing same in proper and competent hands to put down enough wells to thoroughly test the oil and gas possibilities of the state's property. If the Legislature can spend millions in building an asylum for the insane and a half million for ticks, I am sure that the people would approve the expenditure of this relatively small amount to test properly the oil and gas prospects on their own land. In making this suggestion I trust that, no one will charge that I am advocating that the state go into the "wild-cat" business, but with recent developments such an adventure would no longer be considered a "wild-cat" scheme.

Governor Bilbo addressed the legislature again in April 1930 with regard to his veto of a conservation bill:

I am returning herewith House Bill No. 380, "An Act regulating the drilling of natural gas and oil wells; regulating the conservation of gas and oil; providing for the supervision of gas and oil operations by the State Geologist and the Mississippi Railroad Commission; and fixing the penalties for the violation hereof" without my approval for the following reasons:

I am firmly of the opinion that the State of Mississippi at this time is not ready for such a strict law regulating the conservation of gas and oil. I can see no necessity to regulate something we do not have in any appreciable commercial quantities. There is not within the confines of the State a definitely proven oil or gas area where production is sufficient to warrant such strict statutory regulation.

It is a well known fact that practically all the oil fields of the country were discovered and brought in through the efforts of independent operators. The big oil companies have never been, in recent years especially, interested in the discovery and development of additional oil and gas bearing territories and for the reason that there is at present an over production in existing fields of both oil and gas. It therefore follows that the chief hope of the people of Mississippi for the discovery of oil and gas in this State lies unmistakably in the drilling activities of the independent operators. No conservation law worth while can be enacted that will not restrict, hinder and retard the operations and proposed developments and multiplied activities of these independent operators. A conservation law at this time would be to a very marked degree the restriction of both oil and gas development in Mississippi to the big operators, the very class of men who for reasons already stated are not desirous now of increasing oil and gas production anywhere in the world. What Mississippi needs is to give to every man, whomsoever he may be, the right and full freedom to lay his hand to the drill and sink wells upon every hill and in every valley in the State where the geological structure indicates any possibility of successful drilling.

Jackson's Natural Gas Franchise

With new gas wells being completed almost daily around Jackson, the operators were anxious to sell gas. The Jackson City Commission appeared reluctant to rush into a franchise arrangement and held out for a rate equal to that of the city of Monroe, which was 30 cents per 1,000 cubic feet (MCF) of gas for domestic use delivered to the customer. After the Mendoza well came in, Jackson Mayor Walter Scott rejected a proposal from the officials of Love Petroleum Company to supply the city. Becoming impatient, Big Boy Love made a deal with Mississippi Power & Light Company to supply its electrical-power generating station with two million cubic feet of gas per day from the Mendoza well for a six-month period.

The Pearl River flooded and delayed the building of a line to the power plant until midyear, but a four-inch welded line costing $8,000 began supplying gas on June 9, 1930. Mississippi Power & Light Company paid a flat monthly rate of $3,250 for whatever amount of gas it used, averaging some two million cubic feet of gas per day.

City and state officials, caught up in the excitement of the gas wells at Jackson, believed that the state was on the threshold of becoming a major oil and gas producer. Mayor Scott issued the following public letter:

> With eight wells now [capable of] flowing approximately TWO HUNDRED MILLION cubic feet of natural gas per day and the fact that some twenty other wells are being drilled in Jackson, there no longer remains a doubt but that our State has a natural gas field that will equal if not surpass any gas field in the United States.
>
> The City Officials of Jackson are making thorough study and are carefully considering proposed franchises and rates with an abundance of natural gas within our city limits and under our streets and under city owned lands, we are contending for rates equally as low as in any other city. We believe it is for the best interest of not only our city but for the entire state that all Mississippi municipalities should co-operate in the development and the use of Mississippi natural gas.

In August 1930 Scott directed the city engineer to complete plans for piping the city with natural gas and to design a system that would supply a town of 100,000 people, twice the 1930 population. Apparently, the cost appeared prohibitive. Instead, the city commissioners favored a proposal by Mississippi Power & Light, which had acquired the artificial gas franchise through a merger with Jackson Gas Light Company in 1923, and had a pipe distribution system in place to many homes and businesses in Jackson. Harvey C. Couch, president of MP&L, along with vice president Louis V. Sutton, made an offer to supply gas at a base rate of 30 cents per thousand cubic feet with various discounts for business, industry, and municipal facilities. The proposal was approved in a city election on October 20, 1930.

United Gas, working with MP&L, made an offer to the producers in the Jackson Gas Field to buy gas at an average price slightly less than four cents per thousand cubic feet. Producers were to be paid on amounts used by the consumers, and would absorb any losses in the distribution system. Natural gas was turned into the MP&L pipeline delivery system for the city on October 30, 1930.

Harry Lambdin

J. H. (Harry) Lambdin, a mechanical engineer with MP&L, was largely in charge of delivering the gas to MP&L's system. At that time, MP&L's power plant was located on South Jefferson Street at a point just south of the present Pearl Street overpass, where the company still has a facility. Immediately behind this was the MP&L Gas Works.

To make the transition from manufactured gas to natural gas, it was necessary to adjust the stoves and heaters throughout the city to accommodate the higher BTU gas. Also, 150 to 175 miles of new pipelines were needed to handle both the low- and high-pressure systems required to serve the city's 50,000 inhabitants and its industries. Many houses and businesses that had not been tied into the manufactured-gas distribution system were now piped to take advantage of the cheap natural gas.

Initially, gas from three Love Petroleum Company wells, the Mayes, the Mendoza, and the Hutton, along with that from Louisiana Gas & Fuel Company well, Toole No. 1, was piped to the MP&L system.

One hazard to be dealt with was that Pearl River occasionally flooded the entire area up to the GM&O Railroad track behind the Old Capitol. Lambdin reports that he often would ferry workmen by boat to each well to spend the night on the "XMAS Tree" above the water level. When the rising water extinguished the heaters that kept the "regulators" from freezing, the regulators had to be operator by hand to keep the wells producing.

Having a market assured, the drilling pace in Jackson accelerated, with as many as 25 rigs running.

Mississippi's First Oil Well

The first oil well to be completed in Mississippi came roaring in on August 4, 1932. This well was in the Jackson Gas Field along Old Highway 49, south of the railroad underpass near the present-day Richland community. The Brad W. Hensley-Warner No. 1 flowed at the rate of 15 barrels per hour of 14-degree gravity oil. The flow was turned into an earthen storage pit.

After a few days production, the well began to make considerable salt water along with the oil. Nonetheless, the well produced off and on for a number of years. Two other wells were drilled nearby, but all three suffered the same fate, making considerable water with the oil. It is estimated that no more than 20,000 barrels of oil were cumulatively produced from these wells during their lifetime. A number of other wells appeared to be oil wells on initial test, but were drowned by saltwater after a few hours.

It was eventually realized that a thin layer of very heavy oil was sandwiched between the gas and an ocean of saltwater below. This layer was probably no more than ten feet in thickness. The oil was as "viscous as pitch" it could not be poured out of a bucket if the weather was cool. Its very high viscosity resisted flow into the well bore and allowed the freely flowing water to break through the oil layer and "water out" the well.

The quality and value of crude oil are usually measured as "gravity." "High gravity" is a very light-weight oil with a large amount of gasoline (naphtha), diesel, and other "lighter ends." High gravities generally range from 32 to 60 degrees. The higher the gravity, the lighter the color and the thinner the oil.

Medium gravity is generally in the range of 20 to 32 degrees. This oil has more "heavies" and is usually dark and thicker.

Low-gravity oil ranges from 12 to 20 degrees and is very heavy, thick, black oil. It contains a large amount of asphalt. On the surface a very low-gravity oil may solidify and must be heated to flow.

The ideal oil is 40 degrees gravity, and it receives the highest premium. The price paid for crude is reduced approximately one percent per degree of gravity, both upwards and downwards from 40 degrees gravity.

The crude oil in the Jackson Gas Field had too low a gravity to be commercial under the circumstances.

B. B. Jones Returns

When the Jackson Gas Field was first discovered in early 1930, B. B. Jones returned to Jackson and began buying leases and minerals for himself and Tom Slick. On August 18, Slick, often called king of the wildcatters, died of a stroke at age 46. At the time of death, the Slick estate was valued at over $35 million. Jones continued his activity with his brother.

Highly respected and well known for his philanthropies, Jones supplied the money to drill a well on the Belhaven College Campus, known as the Belhaven Campus-Fee No. 1. The school owned the well, but Jones would recover his money from gas sold. United Gas followed his example by drilling a well for Millsaps College on its campus.

Jones went on to participate in much of the drilling in Mississippi during the 1930s and eventually became a partner in Jones-O'Brien, a company headquartered in Shreveport. Although not a resident, he spent much time in Mississippi, particularly during the Jackson gas boom and Tinsley period.

He died in 1953. His nephew, E. B. McGehee, participated in oil ventures in Mississippi for many years after Jones's death.

Crystal Lake

One of Cleve Love's wells drilled in 1930, the No. 1 Ridgway-McGehee, flowed saltwater, with very little gas. An enterprising group decided to capitalize on Cleve Love's misfortune. They constructed a swimming pool with wooden sides and a sand floor, which was filled by the flow of hot saltwater from the well. Known as Crystal Lake, it was enjoyed by many Jacksonians for its balmy waters, particularly in cool weather.

The pool was located in the wooded Pearl River bottom along a dirt road extension of High Street. The present-day location would be just north of the Coliseum on the north side of High Street at its intersection with Greymont Street.

Some customers complained of the hot water in warm weather, the fumes, and the salt's burning children's eyes. These complaints were overlooked to a large degree, however, as the enterprising group who built the facility cleverly provided a common shower room for girls and boys. The sexes were only separated by a partition which reached slightly above head level. This prompted some secret peeping, but helped assure a steady crowd of young patrons (Lloyd, 1989).

The Crystal operated for several years, but was eventually closed down and the pool dismantled.

Jackson's Black Oil Man

At least one black citizen of Mississippi participated in the drilling boom in Jackson. Dr. S. D. Redmond, well known for his former position as Republican committeeman, built a derrick to drill a well in Rankin County, but lightning destroyed the derrick. He turned his lease over to Pioneer Oil and Gas Company, which drilled a 53 million-cubic-foot-per-day gas well, completed in November 1930. Redmond went on to become involved in four more producing wells in the field.

W. E. Willis

A more controversial operator in the Jackson Gas Field was W. E. Willis, long a favorite target of Robert Steffey. Steffey alluded in his June 16, 1929, report to "W. E. Willis, the notorious promoter with a long history in Mississippi."

Stop and Think What a Few Dollars Invested in This Company May Mean To You

Small Investments in Oil Have Meant Independence to Thousands

An Afternoon Scene at the Oil Fields of the Alabama Gulf Oil and Gas Co. - Inc. Mobile, Ala.

Remember $10.00 will buy an oil lot today, and the deed to each lot guarantees the owner his full share of the profits of all wells drilled on the entire property. Special terms on ten or more lots. This may be your last opportunity.

FILL OUT THE COUPON and mail at once, or wire reservation at our expense, or write for special information.

Alabama Gulf Oil & Gas Co.
INCORPORATED
52 St. Joseph St.   Mobile, Ala.
(OVER)

A 1917 leaflet offering interest for sale in a well near Mobile. The well was dry at a depth of approximately 3,000 feet. (Courtesy of Boyd L. Bailey, Woodstock, Alabama)

ENORMOUS LAKE OF OIL HERE

Is the Opinion of Oil Experts and Geologists

Drillers rapidly Boring Well No. 1. Now at a depth of 1500 feet. Drillers report that indications are equal to those found in the TEXAS and LOUISIANA oil fields

An Afternoon Scene at the Oil Fields of The Alabama Gulf Oil and Gas Co. Inc. Mobile, Ala.

Lots now selling at $10.00 (subject to advance or withdrawal without notice) The deed to each lot guarantees the owner his full share of the profits of all wells drilled on the property.

No time for delay. Fill out the coupon on reverse side and mail at once, or wire reservation at our expence, or write for special information.

Alabama Gulf Oil & Gas Co., Inc.,

52 St. Joseph Street, Mobile, Ala.

References—Any bank or official in Mobile. (over)

Pictured above is the reverse side of the leaflet that is described on preceding page.

Left: Bernard Bryan (B. B.) Jones from Kosciusko was probably the first Mississippian to become wealthy from oil. He and his brother, Montfort, found oil in Oklahoma during the decade 1910–1920. B. B. invested heavily in Mississippi oil drilling and became a great philanthropist.

A festive day in 1921 when drilling began on the #1 Archer well, five miles south of Satartia, Mississippi. Local citizens pooled their money to test a structure reported in U.S.G.S. Bulletin 641-D, 1916, known as the "Eldorado Monocline" in Warren County.

(Upper l to r) Drilling crew of the #1 Archer well: Long, Blackstock, G. W. Adams, Wirt Adams and Stevens. Local citizens: W. High Harris Sam Newman, E. A. Archer, Dr. Crock, L. V. Hamberlin, B. N. Simrall. (Lower l to r) Mrs. L. V. Hamberlin, Mrs. G. W. Adams, Mrs. E. A. Archer, Ruth Russell, Mrs. A. R. Causey, Mrs. Will Beidenharn, Mr. Will Beidenharn, Master B. N. Simrall, Jr., Mrs. B. N. Simrall, Miss Bessie Legg, and Lee V. Russell (behind Mrs. E. A. Archer)

A wildcat in Wilcox County, Alabama, May 3, 1921; the Peachtree Oil Exploration Co. No. 1 well was dry after many months of drilling.

The Louisiana Standard Oil Company Refinery, Baton Rouge, Louisiana, around 1921; operations began in 1909.

A well near Laurel, Mississippi, in the 1920s

Discovery well of the Amory Field (the Amory Petroleum Company, Bony Carter No. 1) came roaring in
on October 6, 1926, and burned for 90 days as a giant
torch. (Reprinted from Mississippi Geological Survey, Bulletin 57, 1943, Photo by O-Steen)

MISSISSIPPI'S FIRST GAS WELL

Amory Petroleum Co.'s No. 1 Carter, drilled by the company into gas production October 5, 1926. Photo by Mr. Jillson.

The No. 1 Carter well on production. (*Oil and Gas Journal,* Tulsa, Oklahoma, Volume 26, No. 36, January 26, 1928)

James P. Evans from Shreveport was an early lease broker in Mississippi from 1925 into the 1940s; he leased the original Amory block.

A tirade in a later 1929 report continued:

> Willis is one of the sore spots we have over here. How he manages to
> get out of the jails he has already been in is wonderful and he has been in
> several. Bad checks and selling stuff twice or three times seem to be the
> offenses he pulls. Men who have been friendly to him in the past are
> giving him up as dangerous and hopeless. New Orleans, Donaldsonville
> and Amite, Louisiana, and Natchez, Miss., are some of the places he has
> been in trouble.
>
> At Amory to where Willis drifted in after the discovery of gas, he began
> the promotion of the Buffalo Oil & Gas Company, Sec. 4-13-18, Monroe
> County, Miss., which reached about 720 feet and is now abandoned en-
> tirely. In an effort to revive operations, a stockholders' meeting was held
> at Aberdeen and I am informed by one of his former associates that the
> records show that $36,000.00 was handled on that deal. One of the major
> companies in that deal at the time, thought that it would help the cause
> along by furnishing better than $2,000.00 to Willis. He had no sooner
> obtained this cash than he jumped into Memphis and put on a party.
> Willis is an element the like of which we are fortunately not afflicted with
> at present.

With the discovery of gas in Jackson, Willis rushed in and acquired a few
leases in Rankin County. He then formed the Southern Petroleum Company
and promoted two successful gas wells.

When the city of Hattiesburg advertised for bids to supply it with gas,
Willis submitted a winning bid and subsequently turned the deal over to a
Texas group to build the pipeline to Hattiesburg. Willis drilled two more
successful gas wells and built a small refinery in Jackson to supply the Highway Department with road oil and asphalt.

Being quite ambitious, Willis also formed a subsidiary company, Southern
Gas Utilities, Inc., which attempted to build a small gas line from Jackson
to Clinton servicing the two colleges, the high school at Clinton, and several homes. This project went into receivership by February 6, 1932. By
1934 Willis's gas wells and the refinery were also in receivership and subsequently sold by the bankruptcy court.

Steffey's report of April 25, 1934 states:

> He [Willis] is no longer located in Mississippi, being up in Illinois where
> he is being tried these days on fraud charges in connection with the sale
> of interest in the refinery and some of the wells he drilled here in the
> Jackson field.

On July 25, 1936, Steffey reports:

> W. E. Willis, an operator formerly in Mississippi and Louisiana, with whom a great many company representatives have since El Dorado days come in contact, died in Detroit, Michigan and was buried in his old hometown, Sullivan, Indiana, a week ago.

State Government Becomes a Gas Producer

After finally winning the lawsuit against Ella Reader the state of Mississippi set about developing the Insane Asylum lands in north Jackson, beginning in 1936. A State Mineral Lease Commission was formed consisting of the governor, attorney general, state land commissioner, state geologist, and the state oil and gas board supervisor. A sum of $115,000 was appropriated by the legislature to use in drilling the property.

The first state well blew out from a charged shallow sand while drilling at 1,095 feet on August 29, 1936, and flowed wildly at a very high rate for 24 hours (Monroe, 1937).

On that fateful day, Henry M. Kendall, a young Jackson attorney, was returning from a squirrel hunt on horseback through what is now Riverside Park. As he approached the state's drilling rig, it suddenly exploded. Drill-pipe shot into the air with a cloud of mud, rocks, and other debris. The wooden derrick began disintegrating and the crew members ran for their lives. The roar was deafening. Soon a crater erupted like a volcano a short distance away from the rig and swallowed one of the boilers. The next day the blowout killed itself, the crater then filling with water and mud. The drilling contractor who suffered this loss was E. R. Owens. When Riverside Park was formed several years later, the crater was still there, occasionally bubbling gas. The city began dumping riprap from road repairs into the hole which was some 20 yards across. It took several years to completely fill the crater.

The state's next well, the Fee No. 2, was drilled to a depth of 5,530 feet. After a year of drilling and testing, at a cost to the state of $45,750.51, the well was abandoned (Monroe, 1937).

Altogether, the state of Mississippi would drill some eight wells, of which five were gas producers. In February 1939 H. M. Morse, the state Oil and Gas Board supervisor, announced that the wells had "paidout" and were showing a profit. Nevertheless, it was hardly the bonanza of Governor Bilbo's prophecy that "the state's share properly safe guarded would soon pay the state's entire bonded indebtedness and even be great enough to defray all the state's expenses and make our state tax free." The wells were almost depleted by the time they "paid out."

Blowouts in Jackson

A number of blowouts occurred during the drilling of the Jackson Gas Field. First, the Ohio Oil Company had a blowout in August 1930, which was successfully brought under control with no injuries.

In June 1932 an explosion occurred on a well with a leaking XMAS tree in Rankin County. Four people were killed, including the well's owners, Laurie and Wayland Attkisson; J. W. McCarroll, the driller; and Bill Taylor, a roughneck. Bill Quigles, production superintendent for United Gas, arrived from the head office in Shreveport and was able to quench the fire, bringing the well under control with only a small crew of men and a winch truck.

In 1932, the Alexander trustee, Ridgway-McGehee No. 1 well, went wild but was successfully controlled. A more serious blowout occurred in August 1938 in Rankin County, flowing wild and releasing an estimated 55 million cubic feet of gas daily into the air. The well caught fire, shooting flames 300 feet into the sky and lighting the whole city and countryside for 11 nights. The legislature approved $20,000 to help fight the fire, but it was brought under control by the Houston Oil Well Fire Fighting Company for $8,995. Red Adair, owner of the company, claims he was never paid.

Ultimate Gas Recovery

Between 1928 and 1937, some 195 wells were drilled on the Jackson structure, resulting in 137 producing gas wells, three noncommercial oil wells, and 55 dry holes. Additional wells were drilled in later years, as earlier wells played out. Eventually, some 210 wells had produced gas in the field, but no more than 113 at any one time, that being in 1934.

The Jackson Gas Field began declining rapidly by 1939, and all wells had ceased to produce commercially by 1954. Ultimately, all wells in the field combined to produce almost 119 billion cubic feet of gas and some 20,000 barrels of oil. One well continued to produce enough gas to supply Belhaven College until 1983.

By nationwide standards Jackson would be considered a fair-sized field, but major gas fields are classed at one trillion cubic feet or more. In comparison, the Monroe Gas Field has produced over seven trillion cubic feet, or 60 times as much as Jackson. Still, the Jackson Gas Field produced 100 times the quantity of gas yielded by the Amory Field in northern Mississippi.

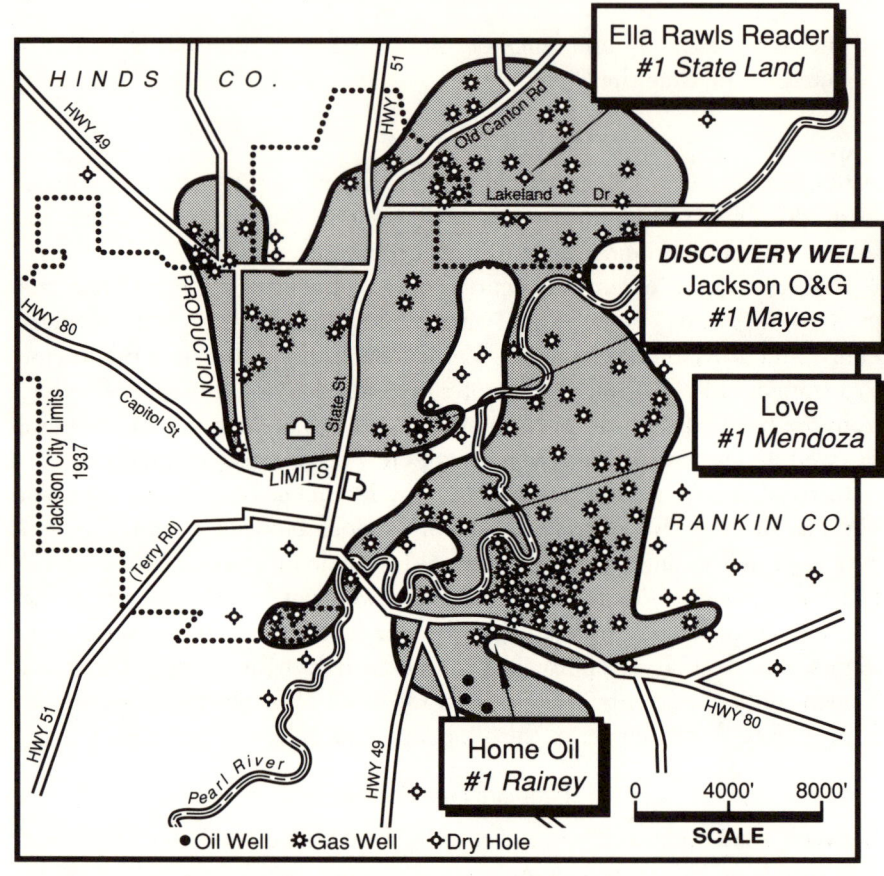

FIG. 5   Jackson Gas Field, 1937 (MSGS Bulletin, 1937, by W. H. Monroe and H. N. Toller)

At the low price of three to four cents per thousand, the total revenue from all the gas produced from the field amounts to only some $5 million. This probably paid for little more than the cost of drilling and completing the wells. However, some operators, such as Love Petroleum Company, no doubt came out very well.

"Over-drilling" contributed a great deal to the financial failure of the project. Many fewer wells could have efficiently drained the structure.

The Jackson Gas field deserves a very important place in the history of the state, more than has generally been accredited to it. The field contributed to the building of a huge pipeline network in the southern portion of Mississippi and Alabama, western Florida, and eastern Louisiana, and supplied gas for this system for ten years without help from other sources.

JACKSON GAS FIELD
HINDS AND RANKIN COUNTIES, MISSISSIPPI
PRODUCTION HISTORY

(Selma Gas Rock)

Year	Producing Wells	Annual Gas (MCF)	Cumulative Gas (MCF)
1930	31	586,634	586,634
1931	68	6,164,370	6,751,004
1932	90	9,674,653	16,415,657
1933	95	9,405,972	25,831,629
1934	113	9,001,229	34,832,858
1935	110	10,290,923	45,234,781
1936	96	12,968,571	58,092,352
1937	91	14,104,293	72,196,645
1938	65	14,050,000	86,246,645
1939	35	15,100,000	101,346,645
1940	26	6,400,000	107,746,645
1941	24	3,950,000	111,696,645
1942	22	2,200,000	113,896,645
1943	18	1,200,000	115,096,645
1944	17	900,000	115,996,645
1945	16	600,000	116,596,645
1946	16	500,000	117,096,645
1947	15	350,000	117,546,645
1948	15	275,000	117,821,645
1949	14	250,000	118,071,645
1950	13	215,000	118,286,645
1951	13	185,000	118,471,645
1952	13	168,000	118,639,645
1953	13	122,000	118,761,645
1954	13	33,000	118,794,645
1955	Remaining wells shut in.		

Development of the Jackson Gas Field and its associated pipelines took place as the Great Depression was engulfing the country. This enabled the city, and to some extent the state, to survive this economic downturn in much better shape that it would have otherwise. The presence of the field pulled oil companies into the state to pour a great deal of money into the economy at a time when it was badly needed. The field encouraged exploration, which was to lead to the big oil discoveries of the 1940s.

CHAPTER 14

Conservation and Tax Laws

With the onset of production from the Jackson Gas Field, the Mississippi legislature once again held lengthy hearings to discuss conservation laws and taxes on the petroleum industry. In 1932, two bills were passed. On June 1, HB660 went into effect, taxing companies and persons "selling or distributing natural or artificial gas to the public" at a flat rate of $25.00 to $400.00 per year, depending on the size of the sales. In addition, on an annual basis, lease brokers were taxed $10.00 per county and pipelines some $10.00 to $50.00 per mile depending on the size of the pipe.

In the same legislative session, SB89 was passed as a conservation bill setting up a state Oil & Gas Board. This became known as the Conservation Act, Chapter 117, Laws of Mississippi, 1932. Dr. George C. Swearingen was appointed by Governor Sonnett Conner as the first state Oil & Gas Board supervisor. On September 23, 1932, the state Oil & Gas Board adopted the first set of rules and regulations, requiring:

- permit, with a fee of $25.00 to drill a well
- accurate logs be kept, available for examination by the state Oil & Gas Board
- casings cemented, to be observed by the Oil & Gas Board supervisor with adequate notice to be present
- standardized pipe size, depth of surface pipe, testing of pipe, and cementing of pipe in the Jackson Gas Field
- notice to the Oil & Gas Board of the intention to abandon a well, with the board to set out plugging procedures

The Oil & Gas Board also was given power to stop an operation and to run a steel line measurement to check depth.

Another set of regulations was adopted to go into effect on January 26, 1933, pertaining to schedule of maximum withdrawals from the Jackson Gas Field wells. At the time it was a common practice for the purchasing companies to produce gas from their own wells at faster rates than other operators from whom they were purchasing gas. The Oil & Gas Board stated that "pipeline companies are pulling the wells unfairly and inequitably and thereby favoring owners of one well against another well."

Under the new regulations well production was to be allowed to be governed according to the well's "open flow capacity"; that is, the maximum rate the well would flow when the valve was wide open. This was reduced by a formula based on the size of the tract in which the well was draining.

Maximum Withdrawal Allowed as a Percent
of Open Flow Capacity

160 acres	24 percent
80 acres	21 "
40 acres	18 "
20 acres	15 "
10 acres	12 "
5 acres	9 "
less than 5 acres	7 "

The minimum allowable was one million cubic feet per day, providing sufficient back pressure was maintained at this rate.

These regulations were challenged in court by United Gas and Love Petroleum Company, both of which had benefitted under the previous arrangement. The legislature followed with statute SB490 in 1936, which gave the state Oil & Gas Board power to prorate wells. A new proration schedule was soon put into effect. Later, in the same year, the federal court overturned Mississippi's proration law. In any event, the Oil & Gas Board began to play a major role in the orderly development of petroleum in the state of Mississippi.

Dr. George C. Swearingen

The first person to hold the position of Oil & Gas Board supervisor in Mississippi was Dr. George C. Swearingen, a native of Georgia. Born in 1866, he became a member of the original faculty at Millsaps College in 1892.

After teaching for 12 years, he resigned in 1904 to become a partner in an insurance firm.

Subsequently, with the discovery of the Jackson Gas Field, Swearingen became interested in geology and drilled a well in the back yard. His residence was at 1501 North State Street across the street from the present location of First Presbyterian Church and adjacent to the Millsaps College Campus.

The well was a roaring success. Henry M. Kendall relates a story concerning this well. It seems that Millsaps at that time had a row of single-room, hastily constructed living quarters known as "the shacks" for divinity students. These were located along Dr. Swearingen's property line. When the well was bailed for completion, the flow line was pointing toward the shacks. The well came in with terrific force, blowing all the windows out of the shacks.

Experience from this incident no doubt helped Dr. Swearingen in his capacity as the first Oil & Gas Board supervisor in 1932. He remained supervisor until his death early in 1936 and was succeeded by Henry Toler.

Attempted Federal Intervention

President Franklin D. Roosevelt formed the National Industrial Recovery Administration as a part of his plan to end the Depression. Harold L. Ickes, appointed administrator of the Oil Section and Code in 1933, almost immediately proposed a moratorium on oil production in view of the tremendous overproduction. Prices had plummeted to 10 cents per barrel. In spite of the bankruptcy threat facing many oil producers, "It sounded to oil men like a crafty path to far reaching government control of the industry" (Tinkle, 1970).

Ickes tried to pass a nationwide conservation bill but was unsuccessful. Nevertheless, he was named Administrator of the Connally "Hot Oil Act," passed in 1935 (Tinkle, 1970).

In order to ward off further federal controls, in 1935 several state governments created an industry-wide conservation program known as the "Interstate Oil Compact Commission." All of the principal producing states became members of the commission. In 1939 President Roosevelt again tried to pass a Petroleum-Conservation bill, but failed to get it by Congress. Each state retained control of its own oil industry. The rules and regulations governing the industry varied widely from state to state.

CHAPTER 15

Capturing the Gas Markets

The discovery of the Jackson Gas Field set off a frantic struggle to capture available gas markets. The principal markets were the numerous southern cities that had been supplied by manufactured gas for the past decades. In addition, large industrial plants, mostly supplied with coal, were prime candidates for natural gas.

Southern Natural

When Southern Natural Gas began building the pipeline from Monroe to Atlanta in 1928 there was no competition from other sources of gas. The manufactured-gas systems were quietly bought out along the route of the pipeline by various outside parties, including the American Commonwealth Power Company of New York and the Central Public Service Corporation of Chicago. All were working with Southern Natural Gas.

These purchases included the manufactured systems at Birmingham, Montgomery, Tuscaloosa, Selma, Anniston, and Decatur in Alabama; Columbus, Meridian, and Hattiesburg in Mississippi; and Atlanta, Macon, Athens, Brunswick, and Decatur in Georgia. It appeared that Southern Natural Gas could capture all of the gas market in the southeastern states at this stage.

In early 1930, the company confidently began constructing a lateral line as an off-shoot of its main line, starting at Benton, Mississippi, and extending to Jackson, Hattiesburg, Mobile, and Pensacola, with several branches to smaller cities. The timing of this project could not have been worse.

The discovery of gas at Jackson, along with the Depression, forced Southern Natural Gas into receivership. The crowning blow came when the franchise was issued to MP&L by the city of Jackson, establishing a base rate of 30 cents per MCF from the Jackson Gas Field wells. In order to finance its new line, Southern Natural Gas was asking for a base rate from $1.00 to $1.75 per MCF, depending on the size of the town.

Having missed out on the natural gas markets at Jackson and Canton, Southern Natural Gas made preparations to supply gas to Hattiesburg, where it already controlled the Mississippi Public Service Company, holder of the manufactured-gas franchise and the distribution system. Southern Natural Gas tied its new pipeline into the Hattiesburg distribution pipes and made preparations to begin delivering gas. Before the gas was turned on, Mayor W. S. F. Tatum, with the backing of the city commissioners, secured a court injunction to prevent Southern Natural Gas from supplying Hattiesburg. The basis of the injunction was that the Mississippi Public Service Company's franchise was to expire on November 7, 1930. Instead, the city commission offered the potential natural gas market to the lowest bidder.

Mississippi Public Service Company submitted a bid with a base rate of $1.35 per MCF. At the same time, the Southern Gas Utilities, Inc. of Jackson, a new start-up company with the notorious W. E. Willis as president, submitted a bid with a base rate of $1.00 per MCF. The commission approved the Willis bid on November 13, 1930, and asked for a referendum vote from the citizens of Hattiesburg on January 20, 1931. The Hattiesburg newspaper strongly recommended approval of the bid submitted by Southern Gas Utilities, and on January 20, voters gave their approval by a vote of 1,409 to 260.

With the loss of the markets at Jackson and Hattiesburg, Southern Natural Gas now tried desperately to capture the principal remaining markets. A spur line was laid to Laurel and tied into the manufactured-gas distribution system. Again opposition arose, and a restraining order prevented the sale of gas. A federal court injunction, however, reversed this restraining order.

Not satisfied, the local American Legion post procured an injunction from the Chancery Court to prevent turning natural gas into the city's system. Again, the federal court overturned the injunction. At last Southern Natural Gas had a customer, and Laurel began receiving gas from its system.

As its pipeline reached Mobile, the company was stopped from supplying gas to that city, again with a fight about rates. Mayor Leon Schwartz submitted the question to the Alabama Public Service Commission, contending that rates being offered could be materially lower by using gas from Jackson,

Mississippi, rather than gas from Monroe, Louisiana, which was being supplied by Southern Natural.

By this time, Southern Natural Gas Corporation was feeling a severe financial pinch and having great difficulty raising new capital.

United Gas

United Gas proved to be the big winner in the scramble to secure gas markets in the coastal areas of the southeastern states. Operating initially as Louisiana Gas and Fuel Company, the company had secured a good lease position in the Jackson Gas Field and drilled several wells. The company already had a large position in the Monroe Gas Field and was the supplier to Southern Natural Gas for 40 percent of the gas shipped to Birmingham and Atlanta.

The United Gas Corporation was incorporated on June 30, 1930, and acted as a holding company for Louisiana Gas and Fuel Company and its other subsidiaries. The company's management recognized the Jackson Gas Field to be a cheap source of gas for the large market in the coastal areas of Mississippi, Alabama, and Florida.

N. C. McGowen, president of the new company, came to Jackson and opened an office in his name. The first employee of the new company was J. R. Rutherford, an engineer, and later Crymes Pittman, an engineer newly graduated from Mississippi A & M was hired. R. S. Stelfox became the company scout and J. S. (Sam) Dodson, the landman. A temporary staff was established in Jackson to conduct negotiations for gas contracts with the producers. Soon United Gas controlled most of the gas in the Jackson Field. The main competition was from Henry L. Doherty of Cities Service.

The producers were offered four cents per thousand MCF for the first five years, escalating to 4.5 cents for the following five years, and five cents thereafter. Crymes Pittman confides, however, that lacking rigid standards, the producers were paid for gas delivered at a higher pressure than that sold to customers, so that in fact, the price being paid was less than four cents per MCF. The gas was purchased at "10 ounces above atmospheric pressure" and sold on the basis of "eight ounces above atmosphere." Company requirements were 30 million cubic feet of gas per day.

United Gas then set out to capture more markets south of Jackson. E. O. Spencer incorporated the Mississippi Industrial Gas Company, and with the backing of Electric Bond and Share and secret backing by United Gas, built a 6 5/8-inch line to McComb.

Pittman, not knowing of the deal between United Gas and Spencer, left his $5.00-a-day job with United Gas to work for Spencer's company for $150.00 a month. When the line was completed in 1931, he was shocked as United Gas took over the project, and all Mississippi Industrial Gas Company employees were fired. However, he was able to go back to work for United Gas and stayed with them until his retirement many years later.

One of the pipeliners involved in the McComb project was Dan Greenwood Hughes. At the age of 21, Dan had been transferred in 1928 from Latex, near Shreveport, to Sterlington, Louisiana, in the heart of the Monroe Gas Field. His employer was Magnolia Gas Company, the gas subsidiary to Magnolia Petroleum Company. With the job came a company house in one of the rows making up the company camp, and, best of all, a company car. He could hardly believe his good fortune. With him was his young wife, the previous Winnie Bell Williams, whom he had met and married in Shreveport. On August 14, 1929, Winnie gave birth to identical twin sons, who were given old family names, Dan and Dudley. Thus this author was born into the oil business.

Magnolia decided to get rid of its low-profit gas company and merged it with the new United Gas System. Suddenly, Dan G. Hughes found himself working for United Gas. With the heavy involvement of United Gas in the Jackson Gas Field, he was sent to Jackson as a company representative during the building of the McComb pipeline. Although the Edwards Hotel near the train station was the principal hangout of oil people in Jackson, Dan chose to stay in Stewart Gammill's plush new Robert E. Lee Hotel near the new Capitol. A few weeks later, he moved his wife and infant sons to an apartment in the city. The apartment had gas for cooking, but heating was by coal, and the twins liked to play in the coal bucket. Thus I became a resident of Jackson for a short time at the early age of one year.

Construction on the pipeline to McComb began in 1930 and was completed in 1931. Dan was transferred next to Summit, Mississippi, and later to McComb during the construction. After the pipeline was finished, he was transferred to Lake Charles, Louisiana, for United Gas Company, and, as the years went on, to Wichita Falls, Palestine, and Dallas, Texas. I did not return to Mississippi until I was 23, in 1953.

To expand its dominance, United Gas Company bought all of the Gulf Refining Company's gas rights in the Jackson Field, including gas wells and some 14,000 acres of leases for a consideration of $500,000. Around the same time, in February 1931, United Gas also bought the Pearl Valley Oil & Gas Company's gas wells and 800 acres of leases in the Jackson Gas Field from Stewart Gammill and Johnny Glassell. United Gas had by now gained control of most of the gas in the Jackson Gas Field and could supply gas at considerably lower rates than Southern Natural Gas could.

United made an offer to buy the Southern Natural Jackson-Mobile line at a distressed price. In desperation, Southern Natural sold the pipeline to United at a loss of $2,387,000. United Gas's purchase included Southern Natural's pipeline from Jackson to Mobile, Alabama, with branch lines to Bogalusa, Louisiana; Canton, Laurel, Hattiesburg, Gulfport, and Biloxi, Mississippi; and the extension of the line under construction to Pensacola. This deal was completed in early 1931. The buying price is believed to have been $1 million (not confirmed). A 15-mile line was laid from the Jackson Gas Field to tie into this system. By February 1931 the Jackson-Mobile line was being supplied by gas from the Jackson Gas Field.

The farsighted actions of United Gas in using the Jackson gas to capture the gas markets in the southeastern coastal area demonstrated the decisions that enabled N. C. McGowen to build a giant gas company in the midst of the Depression.

The money to be made from the Jackson Gas Field was not realized by the producers, who were paid from three to four cents per MCF. The big profit went to pipeline companies and distribution systems, which sold the same gas to customers at $1.00 to $1.75 per MCF. The pipelines continue to operate profitably today, some 50 years later, with gas supplied from newer sources.

Willmut Gas and Oil Company

After the Hattiesburg City Commission awarded the gas franchise to Willis's company, Southern Gas Utilities, Inc., it was necessary for Willis to quickly raise the money to satisfy the $100,000 bond and commence the pipeline.

Supplying natural gas to the city would mean a savings of 75 percent of the cost of manufactured gas. With the help of natural gas engineer, Ralph B. Dudley, Willis interested a group *from Texas* in financing the project. The group formed the Public Service Corporation of Hattiesburg in 1931, with Frank K. McGehee, Harvey B. Barnhout, and B. V. Kincaid, all from Dallas, as officers. This was a separate company from the Mississippi Public Service Company which owned the manfacturers' gas system. Construction on the pipeline began soon afterward, with Mayor Tatum advancing considerable money to the project. Until bonds could be issued and sold, $450,000 was needed to buy the 8 5/8-inch pipe. The 95-mile pipeline was commenced on July 22, 1931, and gas was being furnished to Hattiesburg five months later by November 1931. Willis's wells in the Jackson Gas Field provided the entire supply.

Gulf States Creosoting Company was a large employer at both Jackson and Hattiesburg with two wholly owned gas wells on its Jackson property. The company made a deal with United Gas to transport some of the gas to its plant in Hattiesburg. Mayor Tatum refused to allow United Gas to deliver the gas to the creosote plant, as it would deny market to Willis's new pipeline. In response Gulf States Creosoting Company shut down the Hattiesburg plant, causing mass meetings of the employees, but the mayor refused to budge. Eventually Gulf States went into receivership and sold its wells and drilling rig to Love Petroleum Company.

United Gas made another attempt to take over the gas franchise of Hattiesburg. Through an outside party, the Moran Gas Company from Houston, Texas, United Gas offered to supply Hattiesburg at a base rate of 80 cents per thousand MCF. This was well below the $1.00 rate being charged by the Public Service Corporation. The mayor obtained an injunction to stop Moran from selling gas to Hattiesburg, and an election was called to determine which company was to keep the franchise. On April 12, 1932, the challenger was defeated by a vote of 1,150 to 468. The citizens still supported their mayor, but not as strongly. Conveniently, the Mississippi legislature passed a bill allowing a city government to withhold any such challenge to an existing franchise for a period of two years after such an election. This prompted United Gas to give up its attempt to supply gas to Hattiesburg.

The old Mississippi Public Service Company, which owned the manufactured-gas distribution system in Hattiesburg, was still supplying some of its customers with manufactured gas even though its franchise had expired. Mayor Tatum and his commissioners adopted a resolution on May 18, 1932, giving Mississippi Public Service Company 90 days to remove all of its properties from Hattiesburg. They also required that MPS post a bond of $25,000 to cover damages before the pipelines were removed from the streets and alleys.

A suit was filed against Mayor Tatum and the city commissioners in the federal district court at Biloxi, claiming conflict of interest, but Judge Holmes exonerated Mayor Tatum of any misdealing. His opinion stated:

> All I know is what appears in this case, and he may be the worst man in the State of Mississippi, but there is no proof here of anything that he

has done in this matter that reflects on him in any way. He seems to have been a benefactor in that community so far as bringing cheaper rates to the people down there, and it has not been pointed out to me where he has committed any crime of any character at all, or anything that should disqualify him from continuing to serve as the Mayor of the town of Hattiesburg and to vote on every question that is presented to the council or the Mayor of Hattiesburg, unless the court is going to enjoin the Mayor from acting simply because he, as an official, wants to be just, honest and fair, and wants to operate in the interest of the people. . . .

This case here, as I see it, has no merit in it at all so far as it was presented to me today. It has the refreshing attitude of a public officer looking after the interests of his people, and when he is about to succeed, we have the deplorable state of affairs of an attempt to destroy him because he has stood out for the rights of the people in this regard.

The Mississippi Public Service Company went into receivership and shut down its artificial-gas plant on June 30, 1932.

The Hattiesburg system was being supplied exclusively by the gas wells in the Jackson Gas Field owned by W. E. Willis, who was in financial trouble. Mayor Tatum helped out with a loan of $15,000 in return for a deed of trust on his wells. In September 1932 Willis went into receivership, and the mayor foreclosed on the properties.

By 1933 the Public Service Corporation, which owned the cross-country transmission pipeline from Jackson to Hattiesburg, also went into receivership. The assets were ordered sold by the court on February 12, 1934, including the pipeline to Hattiesburg, with distribution systems at Hattiesburg, Collins, Mt. Olive, Magee, Mendenhall, and D'Lo. Mayor Tatum bought the corporation's assets at the bankruptcy sale at an extremely distressed price.

He formed a new gas company, Willmut Gas and Oil. The name "Willmut" was derived from the first four letters of the mayor's name, "Willie," and the last three letters of "Tatum" spelled backwards. (His imposing full name is Willie Scion Franklin Tatum.) The mayor then placed his sons as officers of the company: Frank M. Tatum as president, W. S. Tatum as secretary/treasurer; and W. O. Tatum as vice president.

The Willmut Gas and Oil Company consisted of four gas wells in the Jackson Gas Field, the 8 5/8-inch, 95-mile pipeline from Jackson to Hattiesburg, and distribution systems at Hattiesburg, Collins, Mt. Olive, Magee, Mendenhall, and D'Lo. Part of the Hattiesburg distribution consisted of the defunct manufactured-gas system. All had been acquired in bankruptcy sales.

By 1936, it appeared that the gas supply for Willmut Gas and Oil Company could run out. The Tatums then proceeded to buy leases and gas wells in the Jackson Gas Field and to drill three successful gas wells plus two dry holes. The company went on to participate in many wildcat wells, including another try at Amory. Willmut did much of its drilling with Cleve Love and later his son, Bill Love.

As the Jackson Gas Field depleted, Willmut Gas and Oil Company was eventually forced to buy gas from United Gas to supply the distribution system. The pipeline to Jackson was later sold to United Gas, but as of 1992 Willmut still supplied the City of Hattiesburg. Willie Scion Franklin Tatum was mayor of Hattiesburg until 1938. He had originally arrived in Hattiesburg in 1893, already in his thirties, and gone into the lumber business. Tatum became one of the wealthiest men in Mississippi, owning many thousands of acres of timberland. He purchased a branch of the GM&O Railroad that ran from Leakesville to Hattiesburg to transport logs to his mill.

A common sight around Hattiesburg was Mayor Tatum in his chauffeur driven Packard limousine. Many stories abound. It has been said that Mayor Tatum was very religious. He meticuously kept accounting records, with a separate column for God in which ten percent, a tithe, of all his earnings were recorded. However, when a tornado tore through his timberlands, he had the damage appraised and charged that loss against God's share (Ladner, 1989). Mayor Tatum died in 1948.

The Canton Gas Pipeline

At the time the ill-fated Southern Natural Gas lateral line from Benton, Mississippi, to Pensacola, Florida, was begun in early 1930, it was contemplated that the line would also supply all of the smaller communities along the way.

The first city to be reached by the line was Canton, Mississippi, and gas was made available to the city in 1930. Even though gas was delivered to the "city gate," Canton refused to accept the gas at the price offered by Southern Natural and held out for rates similar to those which had been agreed to in the Jackson Gas Field. Four years passed, and Canton was still without natural gas. However, Mayor C. N. Harris and the Board of Aldermen applied to the National Recovery Act (NRA) committee in the state to receive a grant from the Public Works Administration (PWA) in Washington to build a city-owned pipeline from the Jackson Gas Field to Canton. This would involve a 24-mile-long pipeline to supply the population of 4,000 in the Canton area.

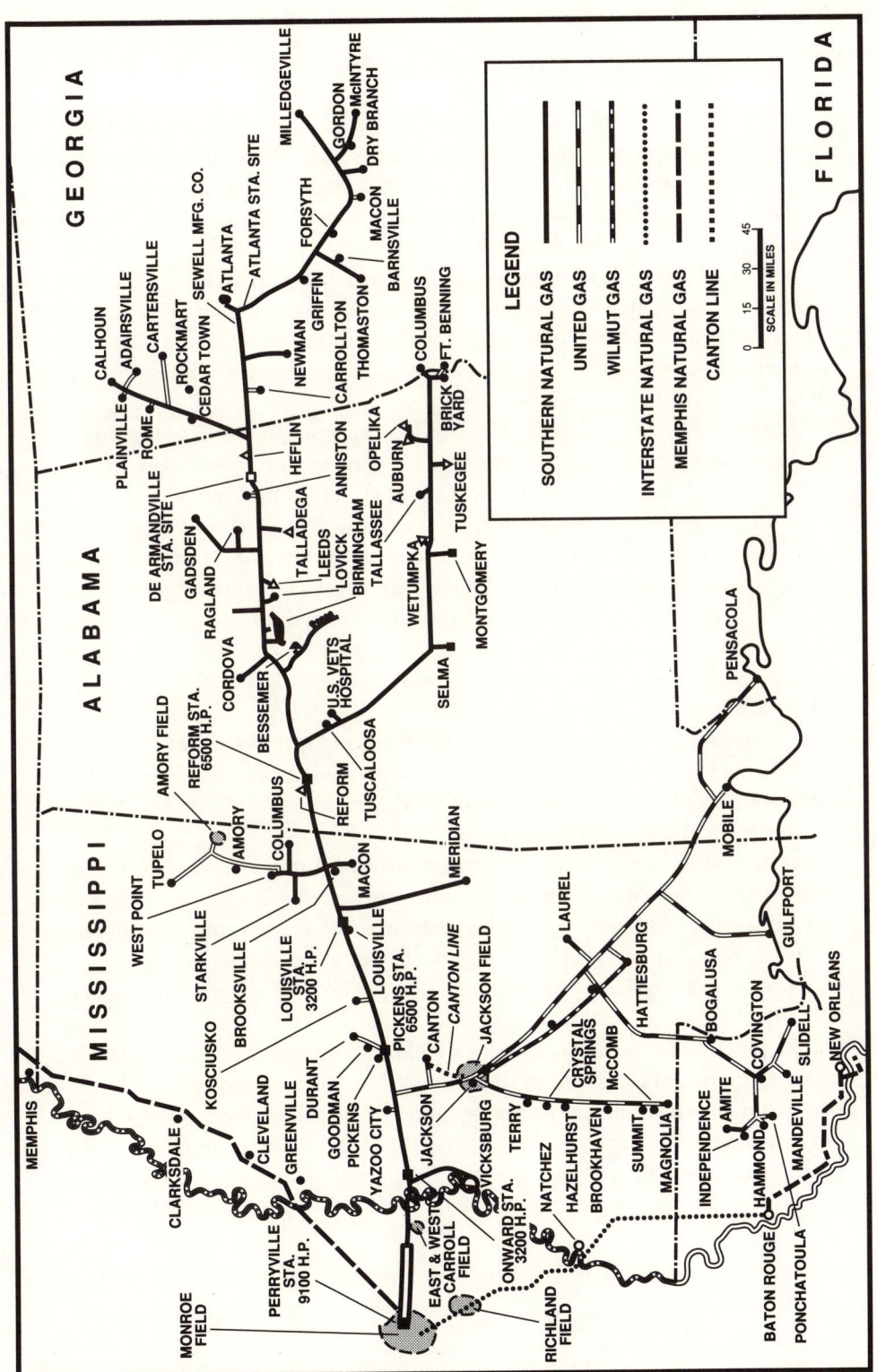

FIG. 6   Natural gas pipeline systems of the Southeastern states, August 1937

The city of Canton took an option on the Belhaven College 90-acre property where B. B. Jones had drilled the good gas well. A special election was held in Canton to ratify the project, with the citizens voting 376 to 40 for the PWA revenue bond issue. Before the PWA grant could be approved, a second gas well had to be obtained.

The city of Canton repaid Jones the $24,000 cost of his well on the Belhaven campus. It then contracted with E. R. Owens, Jackson drilling contractor, to drill a second well for $11,500, with Owens guaranteeing that the well would produce a minimum of 20 million cubic feet of gas per day. The drilling cost had gone down.

Canton's well, the Belhaven College No. 2, spudded October 31, 1934. After 26 days of drilling, the well was completed on November 26 for 27,124,000 cubic feet of gas per day. Owens collected his money.

Having fulfilled the obligation for a second well, money was granted by the PWA amounting to $45,000 in an outright grant and $145,000 being financed on four-percent revenue bonds. Construction commenced in April 1935 on a 6 5/8-inch steel welded pipeline. The right-of-way was along Highway 51 for some 25 miles to Canton. Four-inch cast-iron pipe was to be used for the distribution system within the city.

The PWA grant required that contractors pay 40 cents per hour to common labor and $1.00 an hour to skilled labor, and only local help could be employed. However, it was necessary to hire experienced welders from out of the state due to the shortage of skill in Mississippi. Two shifts of 25 laborers were used in the construction.

By the end of 1935 gas was being supplied to Canton at rates lower than those of any other city in Mississippi. The city hoped that this would attract new industry. A January 1936 report indicated the Canton pipeline project was using approximately 14 million cubic feet of gas per month.

The city of Vicksburg, which was being supplied by Southern Natural Gas from the Monroe gas field, also applied to the PWA for a similar grant to build a pipeline from the Jackson Gas Field. A contract was made with the state of Mississippi for gas from the wells on the state Insane Asylum lands. Even though Mayor J. C. Hamilton and the commissioners received approval from the Vicksburg voters, a snag developed in Washington and the project was rejected.

CHAPTER 16

Developments in the 1930s

Oil at Quitman

The Gulf Refining Company continued to explore relentlessly for oil and gas in Mississippi and Alabama and even making some inroads into Florida. Since 1926, Gulf had been building a large databank of surface-geology maps, magnetometry work, gravity surveys, core-drill information and, by 1931, reflection seismograph. Gulf also drilled many wells and participated in other operators' wells. Its only success to date had been a few wells in the Jackson Gas Field, which had been sold to United Gas.

One of the areas that had appeared promising on the 1929 surface mapping was near the city of Quitman, Mississippi, where faults had been mapped. Beginning in early 1931, Gulf drilled some 23 core holes to further define the structure. Gulf then took a 25,000-acre lease block, primarily from Long Bell Lumber Company, with the obligation to drill two 3,000-foot test wells. Involved in this project were Gulf's landman, Downer, and geologist, Dorchester. The first well drilled in May 1931 was dry. A second well drilled in the fall of the same year encountered good oil shows in the Eutaw sand. Seven-inch pipe was set at 3,685 feet.

Tremendous excitement was created by this prospective oil discovery. Steffey reported, "It is generally conceived that this is one of the most important discoveries that has yet developed in the State of Mississippi." Royalty began trading at $20.00 to $25.00 per acre. Oil people rushed to the area filling up the hotels at Meridian, Quitman, and Laurel. Altogether, approximately 300 oilmen witnessed the testing of the well. The oil crowds in these hotels could get quite rowdy at night, with bottles of bourbon materializing out of suitcases, even though Mississippi was a dry state. Sometimes all-night vigils took place as the men awaited the results of tests on a well.

The well failed to flow. It was "put on pump" to settle out at some 15 barrels a day of 15-degree oil, plus 90 percent saltwater. Although not commercial, it was very encouraging that oil was being produced in Mississippi, 100 miles farther east than the minor oil production in the Jackson Gas Field. The well was deepened and then abandoned. So far only "heavy" oil had been found in Mississippi.

One of the men who came to Quitman was J. E. Stack with his 17-year-old son, J. E. Stack, Jr. Young Stack was indelibly impressed with the oil being swabbed. Many years later he drilled several additional shallow wells at Quitman, making only small unprofitable pumpers. Finally, 35 years after everyone else had given up, he would find a prolific field at greater depths. He never doubted it was there.

The Conroe Trend

In 1931 the giant Conroe Field was found north of Houston, Texas, which would ultimately produced some 600 million barrels of oil. The formation from which this field produced, the Cockfield, could be correlated with similar sands in Mississippi, and the theory was developed that the "Conroe Trend" might extend through Louisiana into southern Mississippi and possibly on into Alabama, the Florida Panhandle, and western Georgia.

When this theory became popular in 1932–33, Gulf and United Gas began concentrating their exploration efforts in the extreme southern portion of the state. Gulf leased 1.5 million acres along the coastal counties, tying up almost all the large lumber-company tracts, while United concentrated in Wilkinson, Greene, and George counties. A number of wells were drilled by both companies, all dry. The trend actually proved to be further south through the Baton Rouge area.

Checkerboard Leasing

A system of taking ten-year scattered leases had developed in the mid-continental areas of Oklahoma, Kansas, and the Texas Panhandle around 1916–17. The "assumption was that other operators would eventually demonstrate the productivity of many pieces which had been bought cheaply" (Owens, 1975). Millions of acres were leased by the large oil companies in Oklahoma, Kansas, and Texas, including Prairie Oil & Gas, Magnolia Petroleum, and the Texas Company, but the leader was Carter Oil Company.

Homer P. Lee organized a gang that did most of Carter's leasing. He estimated his group was buying a lease per minute for months. Carter ended up with 20 million acres in these mid-continental states. Beginning in 1929 and continuing through the 1930s, Lee and his son, Homer P. Lee, Jr., of Tulsa and Dallas, moved into Mississippi. Thousands of acres of leases were bought for Sun and other companies around Jackson and farther south in Amite, Wilkinson, and other Mississippi counties. The Lees also established a royalty company and obtained mineral rights under much of the same land. Since that time Homer P. Lee entities have had large mineral ownership in Mississippi, and often must be dealt with when a block is being leased.

Another checkerboarder who would end up a permanent resident of Mississippi was J. C. (Jack) Vaughn. Beginning about 1931, working for Sun as a lease broker, Vaughn established a network of spies throughout Mississippi who would tip him off when another company was buying leases in an area. He would rush to the area and "bust their block," leaving Sun with a scattering of leases throughout almost every major lease play. Sun subsequently ended up with a number of wells in almost every large field found by other oil companies. Sun was to become known as "the edge oil company" (Storey, 1990).

Alfred S. Black, originally from New York, created considerable problems in 1931, when he took some 600,000 acres of leases through southeastern Mississippi and western Alabama, with his own special lease form. One clause provided,

> this lease shall never be forfeited or cancelled for failure to perform in whole or in part any of its expressed or implied convenants, conditions, or stipulations until it shall have first been finally judicially determined that such failure exists, and after such final determination, lessee is given ninety (90) days to comply.

This could be construed to mean that a lease would never expire on its own terms but could only be released by court action. The clause clouded the titles of many leases and forced companies desiring to drill in the areas to deal with Black.

Black was a "capitalist with a pine stump by-product plant at New Augusta, Mississippi" (Steffey Report, August 15, 1932). He also became a large mineral holder in the areas under lease.

The First Electric Log

In January 1934 E. B. Germany and Tom B. Cranfield, prominent oil operators from Dallas, appeared on the Mississippi oil scene by taking a 20,000-acre lease from Scanlan Lumber Company in Lamar County near Hattiesburg. A. F. Crider, geologist from Shreveport, was employed to do geological work and Si Borden was contracted to drill a well.

The well, the J. J. Newman Lumber Company No. 1, was spudded August 15, 1934, and abandoned as a dry hole September 11, 1934, at a depth of 3,520 feet. The well was reported to be running from "200 to 800 feet higher than regional dip" (Steffey).

It had become a common practice to run a wire-line measuring device at total depth to check the accuracy of the manual pipe measurements performed by the drilling crew. Germany chose to use Schlumberger Well Surveying Corporation of Houston, a relatively new company that not only measured depth but also obtained electrical readings of the formations. These readings tended to show the lithology of the beds through which the hole was drilled, in addition to the likelihood of hydrocarbons or saltwater being present in the porous zones. This appears to be the first well in which an "electric log" was run in the southeastern states. The log was so successful that from the time of this well forward, almost every well would run an electric log.

Conrad and Marcel Schlumberger had developed the logging tool in France in 1927 and had established the Houston company in 1934, the same year this logging was done.

A Salt Dome is Discovered in Mississippi

Encouraged by the high structural position of the first well, Germany began core drilling the area in late 1934, apparently in conjunction with Sun Oil. To clear title to their 20,000-acre Scanlan Lumber Company lease, Germany and Cranfield paid back taxes of 91 cents per acre, and in exchange received seven-eighths the minerals.

Sun Oil bought a 50 percent interest in the deal in April 1935 for $10,000 and took over as operator. Core drilling was continued for several weeks. At least 11 holes were drilled. The Steffey report of March 10, 1938, credits Sun's geologist, Don Monroe, with doing the surface mapping which originally located this structure, thereby explaining Sun's involvement. Germany and Cranfield sold Magnolia Petroleum Company, as well as several other parties, portions of their interest.

A new well was started during September 1936. The Sun Oil, Scanlan, & Semmes No. 1 surprised everyone by encountering anhydrite cap rock at 1,655 feet and salt at 2,522 feet. The well was carried on to 4,024 feet, staying in salt for the last 1,500 feet with no oil found. Sun's well was plugged January 12, 1937.

The Gulf of Mexico had first subsided below sea level and filled with Atlantic Ocean water some 200 million years ago. Extremely arid climatic conditions prevailed throughout the region, unlike today's subtropical climate. This "Sahara desert"-type climate caused the waters in the gulf to evaporate and leave a thick layer of salt for hundreds of miles reaching up to several thousand feet in thickness. This salt layer is now buried from two to six miles deep by sediments that filled in the gulf as the basin resumed its subsidence.

Salt is a crystalline hard rock on the surface. However, under heat and pressure caused by the weight of overlying rock, it acts as a very thick fluid that slowly creeps to points of least resistance. Salt is also of lighter weight than overlying rock, which often causes the salt to try to rise to the surface. The unstable salt layer in the Gulf Coastal Basin has moved about over millions of years to form structures that arch and fault the overlying beds. These structures became traps for oil and gas. "Piercement salt domes," averaging one mile in diameter, actually penetrate overlying beds, and extend as giant vertical columns several miles upward from the mother salt layer to within a few thousand or even few hundred feet of the surface. Larger "deep-seated domes" affect a much broader area. These form huge anticlines over the salt that can hold millions of barrels of oil.

Most of the Gulf coastal oil and gas fields of Texas and Louisiana were a result of salt-generated structures. Speculation held that salt structures might also exist in the southeastern states. Sun Oil's discovery of a salt dome in Mississippi in 1936 confirmed the presence of salt structures, which greatly enhanced the potential of the state as a petroleum province. The salt dome was named the Midway Dome. Sun went on to drill two more shallow wells to the salt, but finding no commercial oil production in the beds arched above the dome, the company elected to drill the fourth well, the Talley No. 4, as a flank test. It was hoped that oil would be found trapped in the upturned beds around the periphery of the salt. But after some five months of drilling, the Talley No. 4 was abandoned in salt at 8,673 feet on February 12, 1938. This well set a new depth record for the state.

Mississippi's Second Salt Dome

After emerging from receivership in 1936 Southern Natural Gas began exploring for gas. The first venture, originated by the new chief geologist, Geoffrey Jeffreys, was to take an option on a 23,000-acre block from R. R. Chichester in western Hinds and Warren counties for the sum of $5,000. The area had originally been surface-mapped by Henry N. Toler who had found a structural high. The company moved in a Petty seismic crew and began a geophysical survey of the area. Jeffreys correctly identified the structure to be a salt dome.

After several months of shooting, Southern Natural Gas began drilling its Angelo-Williams No. 1 well in late 1937, only to find Mississippi's second salt dome at relatively shallow depths. Cap rock was hit at 2,775 feet and salt at 3,026 feet. The well was drilled on to 4,346 feet in salt to establish the fact that it was a piercement salt dome. The well was dry and abandoned December 15, 1937.

This second dome in Mississippi was named the Edwards Salt Dome. Southern Natural Gas drilled a flank test on the dome in 1938, to a depth of 8,458 feet with no success. The Edwards Dome was located some 100 miles northwest of the Midway Dome. Now geologists realized that Mississippi was truly underlain by an extensive salt layer. There was no doubt that many structures were hidden in the subsurface that could trap oil.

CHAPTER 17

The Oil Fraternity of the 1930s

Southern Natural Rebounds

After being forced into receivership on October 1, 1931, Southern Natural Gas Corporation began restructuring and was ready to resume operations some four years later on January 1, 1936, as Southern Natural Gas Company. Total system capacity was 109,000 MCF per day. By adding additional compressors, the system was soon boosted to 125,000 MCF per day (Ivey, 1967).

Up until 1936, the company had spent relatively small sums in searching for new gas supplies. With the declining pressure of the Monroe and Richton gas fields, the first major exploration activity for new gas was commenced in late 1936.

Geoffrey Jeffreys was hired as new chief geologist in 1936 to be stationed in Jackson. However, he resigned in 1938 to operate on his own. With the departure of Jeffreys, Henry Toler resigned as supervisor of the state Oil and Gas Board to take charge of the geological department of Southern Natural Gas.

Geoffrey Jeffreys

One of the premier geologists of the world, Geoffrey Jeffreys, arrived in the southeastern states in 1936 as chief geologist for Southern Natural Gas Company. Born in London, England, in 1885, he received his geological degree at Camborne School of Mines and Wellington College. On graduation, he was employed by S. Pearson and Son, Lord Cowdray's company. His first assignment from 1905 to 1913 was in Mexico, where he advanced from field geologist to assistant superintendent.

Jeffreys did the surface geology in Mexico that led to the discovery of the Dos Bocas Field in 1908 and the Potrero del Llano Field in 1909. In 1910, Everette Lee DeGolyer made the location for the fourth Potrero del Llano well in the field. This well proved to be the greatest oil well in the world to that time, flowing 100,000 barrels of oil per day and ultimately producing almost 100 million barrels. Even though DeGolyer is given credit for making this famous well location, Jeffreys actually is credited with discovering the field.

From 1913 to 1917, Jeffreys was chief resident geologist for Dutch Shell in Venezuela. He came to the United States, where he worked in Kentucky and Texas, and then returned to Venezuela and went on to several other South American countries.

By 1925 he had moved his consulting office to New York and had his own seismic company operating in Venezuela. With the downturn in business during the Depression, he took the job in 1936 as chief geologist for Southern Natural Gas of Birmingham, Alabama, and moved to Jackson, Mississippi, soon thereafter. After two years in that position, he resigned in 1938, to become a consultant geologist in Jackson until his death in 1953. Steffey reported,

> his severance with the Southern Natural has been a very amicable one . . .
> we are glad to say that Jeffreys is not "severing" his connection with Mississippi and its possibilities; he believes that there is something over here east of the Mississippi River and will undoubtedly be a factor in time to come. . . . Jeffreys will be at the New Orleans AAPG meeting next week with his wife as chief chaperon. He really needs one. . . .
>
> Everyone who has come in contact with him fully appreciates that he is a geologist of the first [order] and knows his "stuff" (Member of the A.A.P.G. since '26). You will undoubtedly be hearing a great deal more about this gentleman's activities. (Steffey, 1938)

James P. Evans, Sr.

James Parham Evans, Sr., first came to Mississippi as a lease broker to buy oil and gas leases for Amory Petroleum Company in 1925. After the Amory Gas Field was discovered, he became very interested in the state. Through the late 1920s and the 1930s, he bought many lease blocks in Mississippi for himself and other companies. Although Evans played a big part in the early oil and gas exploration of the state, he never became a resident of Mississippi, preferring to work from Shreveport, Louisiana.

Born at Austin, Texas, in 1885, he moved to Shreveport, an early age and reached adulthood there. Starting out in the real estate business, he became a full-fledged oil and gas lease broker around 1908. When he arrived in Mississippi during the 1920s, he was already well known in oil and gas circles. After a very successful career, J. P. Evans, Sr., died in 1954.

His son, J. P. Evans, Jr., was born in 1910 and graduated from the University of Oklahoma in 1933, with a degree in geology. Jim Evans, Jr., would become a resident of Mississippi in 1940 and remain in the state as an independent oil operator for the next fifty years, until his death in 1991.

Eastman, Gardiner & Company

One of the largest industries in Laurel was Eastman, Gardiner & Company, a giant lumber company dating back to 1891, with 117,000 acres of land and minerals in southeastern Mississippi. The company, headed by its president Charles Green, occupied a classical Renaissance Italian-style office building in the heart of Laurel. With oil speculation growing in Mississippi, the company decided to develop oil on its own lands.

Initially, small interests were bought in several deals drilled by others. Eastman, Gardiner then hired geologist Arthur C. Trowbridge and his assistant, Urban B. Hughes, to do geological work around Collins, Mississippi. The company's first well was located on its lands near Mt. Olive. Drilling began in December 1932, and the well was completed as a dry hole in February 1933 at a depth of 3,519 feet. A second well was drilled on the edge of Collins but abandoned at 1,685 feet.

Although the first two wells had been drilled under the Eastman, Gardiner name, the company now created the Covington Oil Company with Charles Green, Wallace B. Rogers, and Frank G. Wisner as incorporators. Thereafter, a new company was formed for each well drilled in keeping with the custom of the day.

Eastman, Gardiner next located a well east of Collins, based on a magnetometer survey, and sold an interest to B. B. Jones and Sun Oil Company. The well was drilled by contractors Owens and Modisett to 8,002 feet. This well was dry and abandoned October 15, 1933, but later was found to be on the flank of the Kola Salt Dome. It was the deepest well to have been drilled in Mississippi to that time.

By the end of 1933, Eastman, Gardiner hired geologist Hughes as a full-time employee, later to be assisted by Harry Bey and Bill George. Urban Hughes was to become a famous Mississippi geologist while Harry Bey left Laurel and eventually became district manager for Texaco in New Orleans (Green, Sr., 1987).

After three dry holes, Green was interested in taking a more cautious approach to drillsite locations. The next location was based on a surface anomaly near Waynesboro in Wayne County, which Hughes called his "West King Prospect." Humble had leased a block in this area in 1930, but decided against drilling it. The Eastman, Gardiner well proved to be dry, but narrowly missed the Eutaw oil reservoir of the later-to-be-discovered East Yellow Creek Field by only half a mile.

The company went on to attempt two wells in Clarke County in the Quitman area, both of which were also dry. This discouraged the Eastman, Gardiner group from drilling another prospect which the company had under lease in the vicinity of Heidelberg. Had they drilled the location recommended by Hughes on the Helen Morrison lease, they would have found the giant Heidelberg Field in 1933, ten years before Gulf discovered the field. Heidelberg eventually proved to be a 200-million-barrel oil field:

> It is interesting to note that at this time [1945] Eastman-Gardiner had leased the Heidelberg Area, also on Mr. Hughes' surface work, but decided not to drill Heidelberg due to disappointment at W. King. (Dixie July 19, 1945)

Eastman, Gardiner & Company then proceeded to drill wells in Wilkinson County south of Natchez during 1936 and 1937, which were also unsuccessful. The Eastman, Gardiner & Company became the Green Lumber Company in 1937.

Soon after the Green Lumber Company was formed, the royalty (one-eighth) was separated from the 117,000 acres of mineral rights and permanently assigned to the shareholders. The mineral rights were then assigned to the Smith County Oil Company, which later became the Central Oil Company. This gave the oil company the privilege of granting oil and gas leases on the property, while the royalty was paid to the shareholders. Central was to become one of the largest independent oil companies in the southeastern states during the 1950s and 1960s.

Urban Hughes continued to be the company's principal geologist for the next 25 years. Born in Fulton, Kentucky, in 1889, he received his geological degrees from Vanderbilt and the University of Iowa. His early career began in 1917 with Empire Gas and Fuel Company at Bartlesville, Oklahoma.

Later he worked for Atlantic Refining Company and Bridwell Oil before coming to Laurel in 1929 (Who's Who, 1952).

J. A. Morgan and the Joe Modisett Drilling Company

In 1930 Joe Modisett moved his drilling rig from El Dorado, Arkansas, to Mississippi, and was soon busy drilling wells on a contract basis in the southeastern area. Twenty-year-old J. A. (Jimmy) Morgan of Laurel became a roughneck on the rig and, over the next ten years, moved from job to job in Mississippi, Alabama, Louisiana, and even Florida. He remembered that Modisett and his wife lived on the job in a tent for the first few years. As their financial situation improved, the Modisetts were eventually staying in the finest hotels. Morgan was promoted to more responsible jobs until he was derrick man.

The company began a contract well near Marianna, Florida, in 1932. At a depth of only some 20 feet, the pipe dropped and "returns" were lost. All of the drilling mud disappeared down the hole. Morgan had the drilling crew lower him by rope into the hole until he found himself swinging free in an open cave similar to Carlsbad Cavern in New Mexico. He was pulled up safely. Later, he declared, "as far as I know, I was the only fellow ever stupid enough to go down a hole." Pipe was then set through the opening to the floor of the cave and drilling resumed. A number of drilling rigs were lost in later years when such a cavern caved in to form a "sink hole" that swallowed the rig.

Joe Modisett continued to drill in the southeastern states through the 1930s and eventually originated several drilling blocks, keeping ownership in the wells. His rig was capable of drilling to 8,000 feet so he drilled many of the deeper wells. Modisett died November 24, 1939.

After Modisett's death, Jimmy Morgan became a landman for the Smith County Oil Company, managed by Charles Green, until he entered the army at the beginning of World War II. Morgan became a lifelong Mississippi oil operator after returning to civilian life.

Army Dorchester

One of the best-known oil men in Mississippi during the 1930s was Charles Minor Dorchester, known as "Army." His nickname developed from an earlier working colleague named "Armistead." Their reports to the head office were labeled Armi-Dorchester.

Dorchester spent a great deal of time in Mississippi during the 1920s as a geologist for Gulf Refining Company, working out of the Shreveport office.

As geologist on many of the early wells, he was present at Amory and was "sitting" on the well that discovered the Jackson Gas Field.

As 1930 came to a close, the Depression worsened and Gulf laid off many geologists, Army included. However, he was able to go back to work for Gulf as a landman. In 1932 he was stationed permanently in Jackson, where he lived in the Edwards Hotel, but he kept his residence in Shreveport. This arrangement lasted for many years. Army did much leasing for Gulf and was mentioned often in the Steffey Scout Reports.

By 1937, Gulf was far ahead of most other companies in accumulating information on the Mississippi Interior Salt Basin. A number of very large structures had been mapped. Dorchester was given the task of leasing the more prominent of these during the 1937–38 period. These structures included Eucutta, Heidelberg, and Baxterville, among others.

Dorchester bought blocks of leases for Gulf, but worked with C. R. and L. E. (Preacher) Ridgway and W. E. McGehee to purchase mineral rights under many of the same leases. The group made a deal with B. B. Jones to buy interest in most of the minerals at a promoted price. Altogether, some 20,000 acres of mineral rights were purchased for their group, largely by Preacher Ridgway. The minerals were ideally positioned on geophysical anomalies.

Gulf was buying 10-year leases, calling for one-eighth royalty, at a bonus usually of 25 cents per acre, with 25 cents per acre annual rentals paid to mineral owners, not necessarily surface owners. Following the Gulf transaction, the Ridgway group would offer to buy one-half of the mineral rights at $1.00 per net acre. It was a reasonable gamble that they would get their money back from four years of rentals, providing Gulf did not decide to drop the leases. Years later, when a portion of the mineral rights became very valuable, it appeared to some that the farmers were grossly underpaid. Realistically, at the time a great deal of land in the South was about to be lost to delinquent taxes. The money from mineral and oil lease sales saved many farms.

Julius Ridgway, Preacher's son, stated, "my father would leave the house at 5:00 A.M. and leave for the [Edwards] hotel for coffee at 5:30, then hit the road into whatever county he was working at the time."

Dorchester's business associate, Slew Hester, similarly observes:

> Army appeared at the Edwards House Coffee Shop at 5:00 A.M. daily. This was 30 minutes before opening time. He would sit for 30 minutes and fold napkins for the waitresses who would give him one or more cups

of coffee for his labor [Army was noted for his frugality]. A large group of the oil fraternity would gather at the Edwards House Coffee Shop when it finally opened to discuss prospects and wells that were drilling or were to be drilled (Hester, Jr., 1989).

A good portion of the mineral rights owned by the Dorchester group would later prove to be located in some of the largest oil and gas fields in Mississippi.

The question has arisen many times as to the legality of an employee's buying mineral rights while working for Gulf Refining Company.

In reality, Gulf exhibited a policy of not buying mineral rights for company ownership, only leases, and apparently had a loose policy with regard to its employees' buying mineral rights. When the author worked for Union Producing Company in the 1950s, employees were allowed the privilege of buying royalties and mineral rights, but no leases:

> Army kept a very meticulous file going back to his early days with Gulf. This file . . . contained all of his orders for buying and selling leases. One of the letters ordered him to find interested parties to acquire a mineral interest so that the joint purchase would reduce the lease cost to a minimum by being a combination of a purchase of part of the minerals and a lease at the same time. (Hester, Jr., 1989)

Under conditions existing at the time, Dorchester apparently did nothing illegal, but the magnitude of the purchases may have made some wonder. He worked for Gulf until January 1, 1940, at which time he resigned to become an independent. Up to that time, none of the minerals were in production.

One unconfirmed story is that Gulf gave him the choice of letting the company have the mineral rights or his resignation. Dorchester wisely chose the latter:

> Army often said that the day after he resigned from Gulf, they hired him as a broker and trouble-shooter and by working for them in this category, he made more money in one year on a fee basis than he had made in five years on his salary. (Hester, Jr., 1989)

Gulf, at some point thereafter instigated a policy of not allowing employees to buy minerals and required that all key people sign agreements to that effect.

Steffey paid tribute to Dorchester in his report of December 16, 1939:

Army is one of the most popular and most experienced men in the Oil Business in Mississippi and has certainly been instrumental in putting the Gulf Refining Company into Mississippi in a big way. Mr. Dorchester has been connected with the Gulf Companies for over twenty years.

Fred E. West

One of the best-liked independents of the 1930s was Fred E. West. Impeccably dressed and often wearing a monocle, Fred West was the epitome of a well-groomed gentleman of the 1930s. He was held in the highest esteem by company presidents, southern governors, and oil company employees, even down to roughnecks.

His background was a contrast of events. Greenwood Leflore, great chief of the Choctaw Indians in Mississippi, was his great-great grandfather. Greenwood, Mississippi, and Leflore County were named for the chief. Leflore's large antebellum manor house and plantation were located at Carrol- · ton, Mississippi. Although the Choctaws moved to Oklahoma during 1837–40, the Leflore family, being large landowners, remained in Mississippi.

Fred West had no money for college. On graduation from high school, he became an errand boy for a brokerage company in Memphis, Tennessee. There he met Lord Cowdray (Weetman Pearson) who was doing business with the firm. West was assigned the task of entertaining Pearson during his stay in Memphis. Pearson was impressed with the young man and subsequently offered him the opportunity to travel abroad. For several years West traveled with Pearson to Mexico, South America, and England, becoming a part of the sophisticated world of international business. Pearson also taught him petroleum geology with first hand examples from his wells in Mexico.

After his experience with Pearson, West returned to the United States to graduate from Vanderbilt with a degree in geology. Thereafter, he set himself up as an independent oil operator and promoted a number of wells in Kentucky and Ohio. In the 1920s he sold out for some $200,000. With this small fortune he bought himself a sports car, the famous Apperson Jackrabbit. An accomplished harmonica player, he played while driving around the country in his fashionable automobile, this being before the days of auto radios. With this easygoing way of life he attracted many friendships. Soon West had established homes in several places, including Alabama, Georgia, and Florida. He concentrated on oil deals in the southeastern states.

In 1933, he promoted a five-well deal in the delta area of Mississippi. Westbrook, Thompson, and Stewart, independents from Fort Worth, Texas, agreed to take the deal. Roy Guffey was toolpusher on John Bunn's rig from Ardmore, Oklahoma, which was hired to drill the wells. Guffey's account of the venture:

> The first of these wells was drilled north of Indianola on school land. The second well . . . was on Widow Dodds' plantation, Doddsville, Mississippi. The third well was on Will Dockery's plantation just west of Ruleville, Mississippi. The fourth well was on Sam Knowlton's plantation at Perthshire. . . . The fifth and last well was drilled on the State Prison Farm at Parchman, Mississippi. . . .
>
> [In 1933], Roosevelt closed the dern banks while we were there and nobody had any money except Will Dockery . . . and so he let me have some money to give my crew so they could buy groceries. (Guffey, 1989)

Fred West continued to promote wells in Mississippi during the 1930s, primarily in the delta area.

In the 1930s and 1940s, he negotiated leases on Mobile Bay from the state of Alabama on two occasions. He also leased the Mississippi Sound from the state of Mississippi.

To interest companies in buying these leases West gave a large party at the Buena Vista Hotel in Biloxi, inviting Roy Guffey to be his guest.

> "Pappy, I'm giving a party down in Biloxi. I want you and Mammy to be my guests!" We went into that big, beautiful, old hotel right there off the beach of Biloxi facing the Gulf of Mexico. (Guffey, 1989)

At the party there were many company representatives, including Phillips Petroleum, Kerr-McGee, and the governor of Mississippi. West apparently never made a significant oil find.

Henry N. Toler

One of the young Gulf geologists on a surface-mapping crew working out of Meridian in 1929 was Henry N. Toler. Toler had a geology degree from Louisiana State University with a graduate degree from the University of Illinois. After a brief assignment in South America, and a stint in West Texas, he was transferred to Mississippi by Gulf Refining Company in 1929. There he married his wife, Ruth. Henry suffered the fate of most young geologists on December 31, 1930, when he was laid off, but he elected to stay in Mississippi and was destined to become one of the state's best-known geologists.

Another young Gulf employee who was terminated along with Toler was Corbin D. Fletcher, a paleontologist. The two formed a new consulting firm, Fletcher & Toler, and quickly assembled a block of acreage near Vicksburg, which was sold to a local company at a profit. Soon afterward they were retained to do surface mapping by a large lumber company in western Alabama. While performing this task, tragedy struck. On Good Friday, 1932, the two were having lunch on the porch of a vacant farm house near Riderwood, Alabama, when the roof gave way, falling on Fletcher. Fletcher was taken to the Riderwood Hospital, where he died on Easter Sunday. Steffey reports,

> In spite of the fact that he was one of the World War [I] casualties, Fletcher was able to leave behind better than $30,000 worth of insurance for a widow, a boy of three, and a little girl of seven months. The remains are being taken to Wilmington, Delaware, in the vicinity he will be buried about Wednesday of this week.

Toler was forced to hire another geologist to complete the job for the lumber company. He continued to work for various clients as a consultant, including F. X. Gowans and Alfred C. Glassell.

In early 1933 Toler became a deputy supervisor of the Mississippi state Oil and Gas Board. He went on to become supervisor in 1936 and held that position until he resigned to become chief geologist with Southern Natural Gas in early 1938. Henry Toler continued to be one of the state's most prominent geologists for the next three decades.

Tom McGlothlin

One of the young geologists surface mapping for Gulf in 1929 was Tom McGlothlin, known because of his exuberant personality as "Sparky." McGlothlin reappeared in Mississippi as Gulf's principal southeastern geologist in 1937.

He was born in 1906, as John Thomas McGlothlin. After graduating from Oklahoma University in 1928, he went to work for Gulf as a magnetometer operator. In 1929, he joined the famous Gulf surface-mapping crews in Meridian, Mississippi. After the crews were shut down in 1930, he stayed with Gulf except for a two-year period during 1933 and 1934, when he worked for the Denver & Rio Grande Railroad.

McGlothlin would be the leading Mississippi geologist for Gulf during the company's great oil and gas discoveries of the 1940s. He became a consulting geologist in his later years, working primarily for Masonite.

Fredrick F. Mellen

Fredrick F. Mellen is probably Mississippi's best-known geologist. He was born in 1911 on the campus of Mississippi A&M College (now Mississippi State University), the son of an English professor. Fred was raised on the campus, subjected to much academic influence, and in 1934 received his degree in geology.

After graduation, he worked for TVA in Tennessee for a period before returning to Mississippi A&M College as an instructor. In 1937, Mellen was able to become a geologist for the Mississippi Geological Survey on a WPA finance project. His job was to map surface geology in Winston County, then Yazoo County, and later Warren County.

In October 1938, Fred found an "inlier" of Yazoo clay six miles southeast of Yazoo City that appeared to be structurally higher than normal. Due to the press of time, he went on to work in Warren County. In attempting to write the Yazoo City report, he was puzzled by this phenomenon and investigated further. Returning to the area in February 1939, Mellen found evidence of a large structural uplift near Tinsley (Mellen, 1987).

State Geologist William C. Morse considered this finding significant enough to issue a press release. This would lead to the discovery of the huge Tinsley Field giving Mellen industry-wide recognition. Mellen would spend his entire career in Mississippi and was living in Clinton, at the time of his death in October 1989.

E. B. McGehee

In 1937 Everarde Bernard McGehee moved to Mississippi to enter the oil business. As a nephew of B. B. and Montfort Jones, he had learned the business while working for his uncles in Oklahoma and later in Virginia.

E. B. was born in Memphis in 1905, but graduated from Bristow (Oklahoma) High School in 1924. Following high school, he attended Virginia Military Institute. After working in his Uncle Montfort's bank in Bristow, he was moved to Washington, D.C. in 1930 as an accountant for the Bermont Oil Company. In 1937, he married Imogene Kilgore of Nowater, Oklahoma, and moved to Jackson, where he soon became a prominent figure in oil circles.

The Scouts

Some of the earliest employees of oil companies, even before the turn of the century, were the scouts. The companies depended on information furnished by the scouts to keep up with the drilling, leasing, and activities of other operators.

Before the 1920s, scouts were pretty much on their own swapping information, spying on wells, checking courthouse records, and doing whatever it took to keep their company abreast of oil activity. Geologists relied on scouts to furnish drillers' logs and production information on other companies' wells. The degree of success of a company sometimes depended largely on the abilities of its scouts.

As the oil business spread to many different areas, it was necessary for a company to continually increase the number of scouts to cover all areas of interest. In 1924, M. G. (Buddy) Hale conceived the idea of an oil scout association. The National Oil Scout Association of America was formed in 1925. Over the next few years practically all of the companies with scouts had joined the association. Soon a weekly "scout check" was being held in the main oil centers. At these checks, scouts exchanged information on their own companies' activities. In addition, each scout was assigned a number of counties to scout during the coming week to report any leasing and drilling activities of other operators not represented in the scout check.

With this cooperative effort, a complete scout report could be furnished to the management of each company on a weekly basis at a very reasonable cost. The accurate records on the hundreds of thousands of wells that have been drilled in the United States are a direct result of the scouts, not government agencies. Much of the information would have been lost otherwise.

Beginning in 1930, the National Scout Association of America began publishing an annual yearbook reviewing activities in the oil business throughout the United States.

Smaller companies often assigned geologists or landmen to a dual role, scouting plus their regular duties. Larger companies usually had full-time scouts.

Even though scouts had operated in the southeastern states since the 1920s, it was not until 1937 that the first official scout check was initiated in Hattiesburg, Mississippi. The scouts were W. R. Cox, Texas Company, Hattiesburg, Mississippi; J. C. Hemphill, Magnolia Petroleum Company, New Orleans, Louisiana; J. C. Heyck, Humble Oil & Refining Company, New Orleans, Louisiana; H. L. Johns, Freeport-Sulphur Company, New Orleans, Louisiana; R. E. Stevens, Shell Petroleum Company, Miami, Florida; J. B. Storey, Union Producing Company, Hattiesburg, Mississippi; C. F. Washburn, Stanolind Oil and Gas Company, Hattiesburg,

Mississippi; R. N. Weaver, Sun Oil Company, Hattiesburg, Mississippi. The summary of 1937 was prepared by Henry Toler, state Oil and Gas Board supervisor, and was published in the National Oil Scouts Association of America yearbook of 1938.

The yearbook for 1939 (a summary of activities for 1938) was expanded to include Alabama, Florida, and Georgia with the Mississippi activities. The weekly scout check was moved to Jackson, Mississippi. The active membership increased to include, C. F. Adams, (California Oil Company, Baton Rouge, Louisiana) and Henry N. Toler, (Southern Natural Gas, Jackson, Mississippi).

CHAPTER 18

Other Interesting Events of the 1930s

1931

The Mississippi legislature enacted a $75 million highway program. Hatties-burg Mayor W. S. F. Tatum financed a cement plant project to supply cement for the program.

Steffey reported that C. Frederic Burnaman is the first attorney in Mississippi to specialize in oil and gas law.

1932

Dr. E. N. Lowe, state geologist for Mississippi, issued his 1930–31 Biennial Report which states, "The Jackson Field is regarded as one of the large gas fields of the United States." Dr. Lowe's salary was $225 per month and his assistant's $100 per month.

In view of its new status as a significant producing state, Mississippi bid for the 1933 convention of the American Association of Petroleum Geologists (AAPG), but lost out to another city on the basis that Mississippi's production was not significant enough.

The Jasper County courthouse at Paulding, Mississippi, burned, destroying all records.

1933

W. E. Willis built his small refinery to process the heavy oil in the Jackson Gas Field. He claimed that highways could be blacktopped with asphalt for $3,000 per mile versus $25,000 per mile for concrete.

1934

The salary of the director of the Mississippi Geological Survey was raised to $3,600 per year.

Benedum-Trees of Pittsburgh came back to Mississippi to drill two wildcat wells north of the Jackson Gas Field producing area.

W. E. Willis' heavy-oil refinery at Jackson sold for $750 in a bankruptcy sale. The Jackson Gas Field oil wells produced a total of 1700 barrels of crude oil in 1934.

1935

Clarke County offered a $20,000 reward for anyone completing the first commercial well in the county.

1936

Governor Hugh L. White was inaugurated January 21.

The legislature appropriated $13,860 to the State Geological Survey for a two-year budget. The state Oil and Gas Board received $8,900 for two years, $5,500 (or $2,750 per year) salary for the supervisor, and $2,400 (or $1,200 per year) for office help, with $1,000 for emergencies.

1937

Henry Toler, state oil and gas supervisor, issued a report on Mississippi oil and gas development. He stated that 228 wildcat wells had been drilled, exclusive of wells in the Jackson Gas Field, to September 1937. The four deepest wells were from 7,000 to 8,002 feet.

1938

By the close of 1938, 50 percent of the wells in the Jackson Gas Field had watered out, making it necessary to pull the remaining wells ever harder to meet demand. United Gas was facing disaster if it could not come up with new supplies. The company had its own seismic crews working full time in the search. By now the company had its own rigs.

CHAPTER 19

Alabama, 1926–1938

Whereas Alabama had been the most drilled of the three southeastern states before 1926, the period from 1926 to 1938 was strangely quiet. Figures compiled in 1927 indicated that a total of 158 wells had been drilled in Alabama, but the average depth was only some 1,625 feet with the deepest being 4,130 feet. Many of these wells were located in the Black Warrior Basin, where oil and gas shows were numerous, but where no production had been found except for the small gas fields near Fayette, Jasper, and Huntsville. All of the early shallow gas wells had been abandoned prior to 1926. Dr. Walter B. Jones, state geologist of Alabama, believed that wells had not been drilled deeply enough to have a reasonable chance of producing.

However, two large structural features in Alabama's Gulf Coastal Plain did receive considerable attention: the Hatchetigbee Anticline and the Jackson Fault.

Hatchetigbee Anticline

One of the largest of all structures in the southeastern states is the Hatchetigbee Anticline of Choctaw, Clarke, and Washington counties, Alabama. This huge fold in the earth's crust extends for 30 miles as a long oval shape on geologic maps. Similar structures in some parts of the world contain hundreds of millions or even billions of barrels of oil. The Tombigbee River meanders through its crest exposing inliers of older rock that have been pushed to the surface by some great underlying force.

In the 1850s, the structure was originally recognized by Alabama's first state geologist, Michael Toumey. In 1917, the United States Geological Survey published Bulletin No. 661. by O. B. Hopkins called the "Oil and Gas Possibilities of the Hatchetigbee Anticline." Immediately, it became one of the main focal points of those oil companies that were eyeing the southeastern states as a potential petroleum province.

Several shallow wells had been drilled at various locations on the anticline over the years. Alabama's first well of record was drilled on the anticline in 1884 to a depth of 1,345 feet. In 1902 the St. Stephens Oil Company James Keoughan No. 1 was drilled as a dry hole to 2,006 feet near a landing on the Tombigbee River in Washington County, Alabama. The port has been known as "Oil Well Landing" ever since that time.

It was not until 1928 that the Hatchetigbee became a principal target of the oil company teams. The Steffey Report in late 1928 indicates that old-time independent geologist S. A. Hobson was commencing a well near the old St. Stephens dry hole and, simultaneously, a second well near Frankville, Alabama, on the Hobson brothers' land. Both wells were dry, with only one reaching a sufficient depth to have a chance of producing.

In October 1928 the Gulf Refining Company had seven men doing geological work and buying leases on the Hatchetigbee structure. Gulf acquired 50,000 acres of leases at 50 cents an acre. Humble, Pure, Transcontinental, Ohio, Roxana (Shell), and the Texas Company also had people mapping and buying leases on the giant structure. Most were working out of Jackson, Alabama or Meridian.

Robinson and Greer, independent oil operators from Shreveport, began a wildcat test on the structure three and one-half miles west and slightly south of the village of Souwilpa. Their No. 1 Long Bell-Bolinger well in Choctaw County was spudded in March 1929 with a rotary rig. Brewster drill bits and Elliott core barrels were utilized in the drilling, and two boilers fired by wood furnished the power. The well was dry and abandoned in May 1929 at 4,020 feet after running structurally "low," but with an oil show reported at 4,015 feet in the Eutaw Sand. Magnolia Petroleum, Gulf Refining, and Pure Oil had bought spreads of acreage from the operators to help with the financing of the wildcat.

Following the failure of this well, Gulf decided to core-drill the anticline in July 1929. The dense vegetation and soil cover limited the ability of geologists to accurately locate the crest of the huge 30-mile-long structure by surface mapping alone. Core drilling consisted of drilling a shallow hole to an easily recognizable bed below the surface cover that could be mapped over the whole area. An accurate structure map could then be constructed from these points that would generally reflect the deeper structure where petroleum may be trapped. Gulf paid landowners $1.00 per hole for a drilling permit and left the well for a water well, if the surface pipe was paid for.

The disappointing results of the first Robinson and Greer well did not discourage activity on the Hatchetigbee Anticline. Robinson and Greer announced their intentions to drill two more tests on the structure.

The Texas Company had its landman, J. J. Mason, stationed in Meridian by July 1929, from which base he had taken thousands of acres of mineral rights for the company on the Hatchetigbee. This would prove detrimental to exploration by other operators in future years. The Texas Company, along with some of the other major oil companies that had acquired mineral rights on the structure, would not lease and would likely take a "free ride" on any drilling done by others. The Carter Oil Company became very active on the giant structure and circulated an offer to all its lessors to buy one-half their mineral rights for $1.00 per acre "for the one-half." Magnolia Petroleum Company had three landmen buying leases and minerals on the Hatchetigbee and Jackson faults. Other companies buying acreage were Ramsey Petroleum from Oklahoma City, using landman M. R. Bolinger of Plain Dealing, Louisiana. Bolinger also acquired large mineral spreads for himself. Petro Royalty Company of Tulsa, Oklahoma, was buying leases through its man H. A. Harper; Frank N. Henderson worked individually; Louisiana Oil Refinery had landman Joe Walters working the area; and Ohio Oil Company had Joe Walden on the ground.

Robinson and Greer made the location for a second test, the Douglas Oil Company No. 1, two miles east of Bladen Springs, Alabama. The well was spudded on August 8, 1929. Carter Oil, Sims Oil, Shell, and Gulf Refining all bought spreads around the well from Robinson and Greer to help with the financing. Shortage of cut wood and artesian water flows slowed the drilling. Eight-inch casing was set into the Selma chalk at 2,049 feet to shut off the water flow. The well ran 300 feet structurally higher than the Robinson and Greer first well, but, encountering no oil shows, it was abandoned at 4,028 feet in October 1929.

As 1929 came to a close, companies were still very optimistic about the prospects of finding a major oil accumulation on the Hatchetigbee Anticline, even though five dry holes had been drilled to sufficient depths to test the Eutaw sands along its 30-mile length. Gulf was actively core-drilling and searching for its crest with the fortieth core hole down by January 9, 1930.

With the onset of the Depression, little new drilling was done on the Hatchetigbee until 1937, when Joe Modisett drilled a well to a depth of 7,520 feet, almost 3,000 feet deeper than any previous test. The well was dry, but paleontologist Winnie McGlamery identified the deeper beds to be Lower Cretaceous.

The Jackson Fault

Another very large surface feature, with an aerial length of some 12 miles, lies a few miles southeast of the Hatchetigbee Anticline. This structure had been known to exist since Toumey's time, but was not mapped for the Alabama State Geological Survey until 1905. In that year, Dr. E. A. Smith published a map and cross-section that were the work of S. A. Hobson. Hobson, an attorney and a self-educated geologist, accurately mapped the surface outcrops, but identified the structure as a "dome" or anticline. Twelve years later, Oliver B. Hopkins mapped the same structure for the U.S. Geological Survey and identified the structure as a faulted anticline. Both maps were largely correct. Future drilling and geophysics would eventually prove the structure to be a large faulted anticline caused by a huge underlying "deep seated" salt dome. In geological circles the feature became known as the Jackson Fault.

The Ramsey Petroleum Corporation of Oklahoma City took a large block of leases along the Jackson Fault in the late 1920s, at the same time most of the other companies were concentrating on the Hatchetigbee Anticline. Finally, in September 1931, having delayed drilling due to the Depression, Ramsey Petroleum ran full-page advertisements in several Alabama newspapers inviting citizens with capital to invest in a deep well on the structure. Two of the backers were a Mr. Stapleton of Bay Minette and a Judge Tucker of Grove Hill.

Soon afterward Ramsey made a deal with Danciger Oil and Refining Company to take 50 percent interest in the project and operate the wildcat well. The cost to Ramsey at this time was estimated at $65,000 for leases and overhead. Not surprisingly, the well was located on a tract of land owned by Hobson. After mapping the structure for the Alabama Geological Survey in 1905, Hobson subsequently purchased land and mineral rights on the crest of the structure.

With Danciger's representative, T. S. Stoneman, in charge, drilling began in March 1932. The Tuscaloosa formation was reached in May, and was found to be running an astounding "1,850 feet higher" than regional dip (Steffey Report). This was confirmed by Willard L. Miller, chief geologist for Ramsey, and Doggey Grim from Shreveport, geologist for Tidal Oil Company, later to become Tidewater Oil Company. As soon as this word was out, an army of oil people descended on the area, buying leases and mineral interests. At least 30 of the most prominent Mississippi and Alabama geologists and landmen were present. Hobson took this opportunity to sell one-fifth of his royalty interest under the 40-acre drillsite.

The well reached the depth of 4,534 feet with no oil or gas shows and was abandoned on June 30, 1932. This was a major blow to oil and gas exploration in Alabama, since the state's two most prominent structures, the Hatchetigbee Anticline and the Jackson Fault, had failed to produce.

Other Developments

Over the years, several wells had been drilled between Mobile and Citronelle, Alabama, to shallow depths. One of these wells, just north of Mobile, was making some 30,000 cubic feet of gas daily from a depth of approximately 2,900 feet. This was insufficient flow to be commercial.

Of more potential interest was an old water well at Fort Morgan, an obsolete military post on the peninsula at the southeast end of Mobile Bay. This well, which had been drilled by the U.S. government, had maintained an artesian flow of water for many years. The water was cut with gas, which would burn when exposed to an open flame (and still does today). In spite of this potential hazard, the water was used in the fort's sewage system.

In mid-1932 Danciger Oil and Refining took a lease on a tract from Everett and Boykin, large landowners in the area, with the obligation to drill a 3,500-foot test at Fort Morgan. Perhaps because of their disappointment with the Jackson Fault well, Danciger chose not to begin drilling until January, 1935. At the time, the fort was still in use with a cadre of approximately 1,800 troops. Steffey reported, "The road by way of Foley in Baldwin County will need, in dry weather, a Ford with oversized tires on account of deep sand." The well was drilled to a depth of 2,564 feet, but was dry and abandoned February 19, 1935.

Attention was attracted to the University of Alabama in December 1932 when Professor Arthur Bauder of the Department of Physics and Electrical Engineering announced that he had developed an electrical method of locating oil-bearing strata to a depth of 3,000 feet in the earth. The process consisted of placing two electrical probes on the surface located a mile or so apart and measuring the electrical currents between them. This finding brought a flood of letters with many offers to utilize this process, but Bauder declined all the offers. Although there seemed to be some merit to his process, it was apparently never widely used as a geophysical tool.

In the early 1930s Ohio Oil Company took a block on the west side of Mobile Bay, and subsequently drilled two dry holes. The company moved its long time employee Jerry Walden from Amory to Mobile, where he resided until his retirement in 1936.

The first drilling to any significant depth in the vicinity of Brewton, Alabama, was done in the 1930s. Brewton was later to be the scene of some of the largest oil and gas finds in Alabama. In 1933, Ray L. Eastbrook, with the cooperation of local large landowners headed by Ed Leigh McMillan and other family members of the T. R. Miller Mill Lumber Company, drilled a well six miles south of Brewton. Steffey reported, "A fifty foot heavy timbered dam in a creek less than a hundred yards from the rig furnishes a water supply for drilling operations, swimming pool and bath house for the crew." The well was abandoned at 3,296 feet. Earl McDaniel of West Palm Beach, Florida, put $20,000 into the venture.

Estabrook drilled his second well on T. R. Miller Mill Lumber Company's land in early 1935 with no success. However, the Brewton area continued to be of interest to oil companies, particularly Gulf, as there were surface faults evident in the area. Gulf did considerable core drilling along the fault zone near Brewton.

In June 1935, Mannie McCurrie, a wildcatter from Oklahoma City, established a suite of offices in the Battlehouse Hotel at Mobile. She put on a sales campaign offering leases in a well she planned to drill near Bay Minette, Alabama. The message was broadcast on radio three times a week apparently with success, as United Gas, Arkansas Natural Gas, Sun Oil, Danciger, and many individuals all bought interest in her block at $1.50 per acre, payable at 5,000 feet. With Joe Modisett as contractor, the well was drilled in the summer of 1935 to a depth of 5,026 feet but was dry. However, McCurrie did reach sufficient depth to collect her money.

A hot spring was discovered by J. R. Sealy in October 1927 near Dothan, Alabama, when he drilled a well on his property in search of oil. A steady hot-water flow of some 10,500 gallons per hour was sustained at a temperature of 110 degrees Fahrenheit. Sealy, seeing the possibility of creating a health resort, constructed two large bathhouses with individual quarters for patients. It was advertised that the hot mineral water contained healing medical properties. A restaurant was added and by 1935 the complex was doing a prolific business.

Sealy's investment was estimated to be at least $25,000. Sealy reasoned that the hot mineral water might correlate with the flow encountered at Bladen Springs in Choctaw County, where in the 1800s, several hotels offered similar baths.

The government leaders of Alabama at some point in the 1930s decided the state should get into the oil business and bought leases bordering Mobile Bay in Baldwin County. In early 1936, however, the decision was made that "it was undesirable for the state to be engaged in the oil business," and the state-owned leases were released to the landowners.

At the end of 1938, oil exploration in Alabama was almost at a standstill.

CHAPTER 20

Florida, 1926–1938

Florida received little attention from major oil companies until the 1930s. The state was far removed from the network of pipelines and refineries of the oil states, with marshy terrain over much of the peninsula. It was doubtful that, if oil or gas was found, it could be marketed successfully with the long distances involved. There were few large population centers or industrial areas. The U.S. Geological Survey had placed Florida outside the range of "possible oil-bearing formation" until 1934, when its revised publication placed the entire state in the positive category.

In 1926 a well at Cedar Key that was drilled to some 4,010 feet burned a small amount of gas at the casing head for several months. In 1927, J. L. McCord of Oklahoma reported "sands saturated with crude oil" in a shallow well he drilled at Monticello, just east of Tallahassee. This was never verified.

Robert Steffey began reporting on activities in Florida on a piecemeal basis beginning in October 1928. By 1932 he was issuing a regular report on Florida along with his Mississippi and Alabama reports. All the early activities were by independent operators and promoters. As late as February 1932, Steffey reported, "No major company, as far as we have been able to learn, has yet to put a dollar into oil operations in the state of Florida unless it has been done very indirectly."

Steffey made his first detailed report on a Florida well in October 1928. This was a well drilled to 4,545 feet in Polk County, Florida, at Kissengen Springs. Geologist Claude F. Palmer gave Steffey data on the well including the fact that August Hockschor and Irwin A. Yarnell bore the $100,000 cost. A set of cuttings was furnished to Herman Gunter, the state geologist at Tallahassee.

That same year, Ocala Oil Corporation drilled the No. 1 York well in Marion County, Florida, near the city of Ocala, to a depth of 6,180 feet, which set a new depth record for the state that would not be surpassed until 1939.

Steffey indicated, too, that Gulf Refining Company had 12 magnetometer men working in the Panhandle at De Funiak Springs east of Pensacola in February 1929. And, Thomas A. Edison, visiting Ft. Myers on vacation in May 1931 predicted that those drilling for oil in Florida would either find oil or sulfur similar to that in the Texas coastal fields.

With Steffey issuing reports on Florida, oil companies began to take notice. A well was started south of De Funiak Springs by Ray L. Eastbrook in late 1931, on 182,000 acres leased from the Walton Land and Timber Company. This well, known as the Oil City Corporation No. 1 Walton Landon Timber Company well, was also across the Choctawhatchee Bay from Destin.

It is not surprising, then, that the state of Florida, through its Internal Improvement Board, soon thereafter leased 70,000 acres of the Choctawhatchee Bay to a relatively new company, Islands, Inc., Miami. The Internal Improvement Board also leased 1.3 million acres to Islands, Inc., encompassing all of the tidelands along the west coast of Florida from Choctawhatchee Bay to Key West. These leases covered lands from high tide to a mean water depth of 12 feet. The state leases were granted in January 1932.

Islands, Inc., had been active in Florida for several years. It was owned by a wealthy Washingtonian Francis Whitton, who served as president. William G. Blanchard, a newcomer to the Florida scene originally from Pittsburgh, Pennsylvania, was secretary. He would become the most prominent oil man in Florida during the 1930s.

The state leases carried the obligation to start a well within one year that would be drilled to a depth of not less than 5,000 feet, and further provided that the lessee

> shall not use this lease as the basis of, nor shall he or his successors entered into any general stock selling scheme or plan for the sale of stock to the general public within the state of Florida.

Whitton and Blanchard also drilled a well with resident Florida geologist Dr. Edward A. Hill at Cedar Keys Levy in 1932.

West of Miami on the Tamiami Trail in Dade County, Neil Scroggins had been intermittently drilling a well since 1926. When a hurricane hit Miami on September 18, 1926, the crew crawled into the fire boxes of the boilers in order to save themselves (Steffey, January 28, 1932). The well had reached 4,600 feet by January 1932.

Six miles south of Miami at Coral Gables a well that was known as the "Kendall test" was drilling below 5,200 feet on February 1932. This well was located by a man by the name of Reeber, in much demand because his stomach began to rumble whenever he was near oil. Kiser and Son of Miami was supplying the large amount of cash to drill the well based on a "rumble" at Coral Gable. Altogether, five wells were drilled in Florida in 1932. None were being financed by oil companies, although some geophysical work was being done in the Florida Panhandle. After having been fired as manager of Amory Petroleum Company several years earlier, Judge G. Gillespie had moved to Florida and was drilling a well at Clearwater in 1932.

The first serious exploration of the Florida peninsula was begun in the last part of 1933. Blanchard had leased additional acreage from the State amounting to some two million acres, including the Everglades and Lake Okeechobee. Blanchard was now operating as the Mid-Florida Exploration Company with a new partner, Thomas E. Reedy.

Mid-Florida was headed by Harold S. Foley of Foley, Alabama, and included a group of individuals who had large land holdings, which they united in hopes of attracting oil companies to Florida. Altogether in the pool were some 12 million acres. Included in the group were Burton-Swartz Cypress Company of Perry; Clark-Ray-Johnson from Ocala; Cummer Lumber Company of Jacksonville; Flynn-Harris-Bullard of Jacksonville; Foley's Brooks-Scanlon Corporation; Land Holding Company of Florida at Jacksonville; Putnam Lumber Company of Shamrock; Walter Ray of Jacksonville; Ray Sawmill of Mobile; N. G. Wade Investment Company of Jacksonville; Weaver-Loughridge Lumber Company of Boyd; and Francis S. Whitten of Miami.

Blanchard engaged O. S. Petty, founder and manager of the Petty Geophysical Engineering Company of San Antonio, Texas, to conduct an extensive reflection seismograph survey along a 142-mile line from Ocala to Monticello, Florida. This would entail some 600 shot points. At the same time, a magnetic survey would be run. Petty personally supervised the seismic work from the Miami Biltmore Hotel. The project was commenced on December 5, 1933. A line was also begun by Petty along the Tamiami Trail from Coral Gables to Pinecrest. William M. Barret, Inc., of Shreveport, headed by Randolph H. Mayer, was contracted to do the magnetometer for Blanchard.

It took nine months to complete the geophysical survey after which Blanchard announced that he would begin a "comprehensive drilling campaign" based on results. He had moved the Mid-Florida Exploration offices to Jacksonville. Five oil companies were invited to Jacksonville in May 1934 to review the vast amount of seismograph and magnetometer data that Blanchard had accumulated. The companies sent their geologists, C. L. Moody, Ohio Oil Co., Shreveport; Roy T. Hazzard, Gulf Refining Co., Shreveport; C. C. Clark, United Gas, Houston; George W. Schneider, the Texas Co., Shreveport; and R. B. Whitehead, Atlantic Oil, Dallas. Governor David Scholtz, and Fred C. Elliott of Tallahassee were also present. The governor assured the company representatives that those who came into the state and assisted in the development program "would be looked on most kindly." The visitors toured the area between Tampa and Tallahassee, but made no definite commitments to Blanchard.

Blanchard now devised a grandiose scheme to drill 50 wells on the Florida holdings. The Oil Development Company of Florida was formed with Blanchard as president. A statewide organization known as the Florida Oil League was created and was headed by the leader of the Florida Senate, T. G. Futch. Key leaders all over the state were made vice presidents, and the league was to raise funds to help drill the 50 wells.

In January 1935 Blanchard proposed his first well, known as the South Lake Well in Lake County, Florida. Sam E. Wilson, Jr. of El Dorado, Arkansas, contracted to drill the well to 6,000 feet using his new 120-foot Parkersburg steel derrick, the finest equipment yet to be used in Florida. The well was spudded February 26, 1935, only to encounter an underground cavern and lose returns. The rig had to be skidded 150 feet to try again.

Blanchard drilled the new well to 2,258 feet, at which depth he began coring continuously. A great deal of much needed geological information was gathered. Visiting the well in August were Army Dorchester of Gulf, Ed Lytle of United Gas, Don Monroe of Sun, and others.

Blanchard's well continued to inch slowly downward throughout the year 1935 and on into 1936. In June 1936, 216 core samples from 2,258 to 3,833 feet were turned over to Preston Fergus, division geologist of United Gas out of Monroe, Louisiana, with the understanding that United would furnish the information to other companies. Merle C. Israelsky did the paleontological work for United, followed by further examination of the samples by F. W. Rolshausen with Humble and Mrs. Paul L. Applin.

A dispatch from Blanchard to Steffey on August 3 claimed that a 22-foot core with oil show occurred from 5,000 to 5,022 feet. It also stated that "obnoxious fumes" (no doubt hydrogen sulfide gas) exuded from the core so strongly that they made the onlookers nauseated. Pipe was set and the well cored ahead. By November 1936 the well had been drilling for 21 months and the cost was becoming staggering.

Associated with Blanchard in this well were members of the Arnold family of Groveland, Florida, who were financing the well by selling leases around it for up to $200 per acre. As much as $300,000 was reported to have been raised, probably more than was needed to drill the well to its depth of 6,118 feet. Some of the investors were disgruntled and believed that the deal had been misrepresented. One of the largest investors, the McGraw family of Miami Beach, filed suit against the Arnolds.

Regardless of the method of financing and the legal problems with the well, the wealth of geological information obtained from the cores from 2,258 to 6,118 feet was invaluable to geologists. A Schlumberger survey was run and a completion attempt was made to test the 5,000-foot oil show. No commercial production was found.

In late 1934 the state of Florida decided to conduct its own geophysical investigation of the oil and gas possibilities in the state, and engaged L. Spraragen to conduct a magnetometer survey. The federal government contributed $53,000 through a WPA program toward this project, with the state government contributing $3,000.

United Gas, the only company supplying natural gas in the state (Pensacola area), began a magnetometer survey of the Florida Panhandle area in the summer of 1935 by Jackson S. Young, geologist. Young would later become the United Gas chief geophysicist in Shreveport.

Humble Oil & Refining Company became interested in Florida around 1935. Rather than expose Humble's activities in the state to other companies, CEO Wallace E. Pratt organized Peninsular Oil & Refining Company to act as Humble's agent. The president of Peninsular was geologist Robert B. Campbell. Other officers were Cliff and Wayne Dodge. Humble would absorb Peninsular in 1941.

Interest in Florida now shifted to the Big Cypress area west of the Everglades where Clem S. Clark of Shreveport obtained an option to lease some one million acres in Lee, Collier, and Henry Counties from Barron Collier.

Previously, Clark had shot a seismic line from Miami to Labelle in Henry County. Clark offered this option to several companies. In 1937 he made a deal with the Gulf Oil Corporation of Texas, through the Pittsburgh office, to take over the project. Benedum-Trees, also of Pittsburgh, joined Gulf in the deal for 50 percent. In a newspaper interview, Mike Benedum, wintering in Miami, said "No doubt there is oil in the Everglades." Gulf planned a two-year exploration program to evaluate the lease.

To begin exploration on the Collier lands, Gulf first ran a gravimeter survey utilizing two crews. The swampy land over which the geophysical surveys were being conducted caused Gulf to try a new vehicle, the "swamp buggy," which had been recently developed for use in that type of terrain. The swamp buggy speeded up the surveys to a great extent, but there were problems with many flat tires.

In December 1937 the swamp buggy stalled in the marsh 40 miles from any help. The four-man crew, with J. F. Fair in charge, had to spend the night on the buggy and began the long trek out the next day. They came upon a camp of Seminole Indians where they spent the next night.

Worried about the missing party, Gulf chartered the Goodyear blimp, which was berthed at Miami. The men were finally located 20 miles east of Immokalee. The blimp then took the broken part with the driver to Miami for repairs, while a rescue party carried the other three out to safety on homemade hunting tractors.

The swamp buggy broke an axle in June 1938, forcing the crew to walk 13 miles to the nearest road. This time the blimp, which had come to the rescue on more than one occasion, was off on a trip and could not come to their assistance.

Gulf brought in a seismograph crew in September 1937, and commenced an extensive seismic survey. Following Gulf, Sun Oil Company arrived with a crew in July to do seismic work near Ft. Myers.

With the increased activity in south Florida, the Tamiami Trail became exceedingly popular with oil scouts. Representatives from Gulf, Sun, Shell, Humble, Standard, and the Texas Company were often seen in the area and would manage to spend a few days in Miami.

While the lawsuit raged over his Lake County well, Blanchard moved on to other ventures. He now concentrated on the Ocala Uplift, the most prominent structural feature in Florida. In May 1937 he held a meeting with a group of oil men including Dr. V. R. Garfias, director of Cities Service out of New York; Ira Corn of the Plains Oil Company of Kansas; Gordon L.

Stevens of the Inter Ocean Oil Company; and Judge A. M. Ebright, a prominent oil attorney from Kansas. Nothing came of this meeting.

In a separate venture, Blanchard took a 350,000-acre block at Capes Sable at the extreme lower end of Florida, prompted by the small oil fields that had been found in Cuba. Blanchard and A. C. Preston of Miami were granted preferential leasing rights on all of the state lands in the lower Florida involving over 1.5 million acres.

As 1938 came to a close, the oil companies were finally becoming interested in Florida. Independents by then had drilled over 70 wells in Florida, but only 45 below 1,000 feet, and none had penetrated the depths where production would later be found.

CHAPTER 21

The Stage Is Set

In 1938 Gulf Refining Company stood out as the big player in the south-eastern states. Over the entire 13-year period since 1926, Gulf never let up. First, it had large areas mapped by surface geology crews and magnetometer. This was followed by core-drilling hundreds of holes, particularly in the fault zones. Employing a newly developed technique, gravity maps were made over much of the salt basin. Lastly, seismic surveys were made of many of the gravity features in the salt basin. Seismic surveys were also conducted over broad areas of the coastal zones of Mississippi, Alabama, and Florida.

Along with this costly gathering of information, Gulf leased some 1.5 million acres. A number of lumber companies were able to pay overdue ad valorem taxes on their lands with the lease payments from Gulf, thereby saving their properties. Gulf went on to drill or participate in 40 to 50 wells. None was successful except the minor production the company developed in the Jackson Gas Field and later sold to United Gas.

In 1930 Gulf had drastically cut its staff of Mississippi resident geologists and geophysicists and retrenched to the main office in Shreveport. Army Dorchester, however, reopened a Gulf land office in Jackson in 1932, but other activities continued to be directed from Shreveport. In 1933, G. D. Bilberry, a scout, was moved to Jackson as assistant to Dorchester. Tom McGlothlin joined them around 1937 as geologist.

By the end of 1938 Gulf had leased a number of large structures in the Mississippi Salt Basin based on gravity and reflection seismograph. For 12 years its activities in the southeastern states had heavily drained the company's resources with nothing to show. Unlike other companies, however, Gulf was not deterred, but plunged ahead, still with the conviction that there was big oil to be found east of the Mississippi River. No doubt there was dissension in the managerial ranks as to the merits of this decision.

United Gas

The other big player during the 1926–38 period was United Gas, second only to Gulf. Having contracted for most of the gas in the Jackson Gas Field, United Gas established an eastern gulf coastal pipeline system, which was supplied from this cheap source. The pipeline covered the southern areas of Mississippi and Alabama as well as extreme western Florida and eastern Louisiana. The company's operations in these states soon became profitable, even during the Depression.

In conjunction with its pipeline activities, United Gas managed to establish a fair amount of production in the Jackson Gas Field, some through property purchase from other operators and some through drilling. Realizing that its pipeline system was dependent on this single source of supply, United Gas soon began exploring for new gas supplies.

In 1932 J. Edward Lytle began scouting Mississippi from Monroe and became resident geologist in charge when Jackson Young was transferred back to Monroe in 1934. Lytle was stationed in Hattiesburg. J. B. Storey relieved him in 1937, when Lytle was transferred to New Orleans. At the time, other oil company personnel stationed in Hattiesburg were: Don Monroe, Sun; Carlton Washburn, Stanolind; W. R. Cox, Texaco; Jack Vaughn, Sun; and R. H. Weaver, Sun. Most lived in the Forrest Hotel at $30.00 per month (Storey, 1990).

Beginning in 1937, all the drilling and production of United Gas was transferred to a wholly owned subsidiary, Union Producing Company. During the 1930s, the company drilled and participated in many wells. Notable were several wells in Greene, George, Wilkinson, Issaquenna, and Copiah counties. Several wells were drilled with other operators, particularly Alfred Glassell. There were no discoveries.

The California Company

A third company arrived in Mississippi near the end of the 1926–38 period that was destined to become a big player in the coming decade. This was the California Company, a wholly owned subsidiary of Standard of California. Its activity was confined to geophysical work and buying lease blocks in southwestern Mississippi.

In 1936 the company established a "Salt Dome Oil" group in Houston,

Texas, to do seismic and geologic work in the Gulf Coast. Included in this group were Gage (Bud) Lund, Kenneth Crandall, Hugh Wall, and Fred S. Wright. These men would become the top management of the company in later years.

After successfully shooting seismic in the marsh areas of Texas and Louisiana, the group decided to shoot a line from Lake Pontchartrain to Jackson, Mississippi. In the spring of 1937 the company began the seismic line at Ponchatoula, Louisiana, and worked toward Jackson, using two Geophysical Service, Inc., seismic crews. Geologist J. W. (Soak) Hoover was stationed in Brookhaven, Mississippi, in the Inez Hotel; with him was Robert R. (Bob) Phillips:

> George Cunningham, chief geologist, SOCAL, came through Brookhaven about July 1 [1937]. I was living in *Inez* with John Lewis and Soak. He told Soak that this crew had chosen to run this fast recon to Jackson. (Phillips, December 1989)

At the time, Highway 51 was under construction and traffic north of Brookhaven was detoured to the east side of the highway construction and traffic south of Brookhaven detoured west. To stay out of the way of construction and traffic, the seismic crews ran a cross-country line from Ponchatoula east of Highway 51 and found the Ruth Salt Dome and the Mallalieu structure. On reaching Brookhaven they moved to the west of the highway and found the Brookhaven structure. After completing this work in Mississippi, Hoover and Phillips were sent to Saudia Arabia to do seismic work for The California Company.

Neil Smith ran gravity surveys for the California Company in conjunction with the seismograph, using two Mott-Smith gravity crews.

As the 1930s came to an end, California was doing additional seismic work and buying leases in Lincoln and Franklin counties in Mississippi (Womack, Jr., 1987).

Emmett Vaughey was sent from Houston with his crew of lease brokers to assemble the blocks of leases for The California Company:

> We were located in Jackson. . . . I was most anxious to avoid a scout for the Sun Oil Company called Jack Vaughn. He was well known as being able to hear a lot of information and I did not want him breaking our block. There was one country store in Ruth that sold the farmers everything, clothes, groceries, and what have you. It was run and owned by a man by the name of C. C. Clark. He turned out to be a very good friend

of ours and kept our identity hidden from anybody around and did not gossip about the leases purchased. One time he was visited by the same Jim Vaughn who asked if there was any activity going on and Mr. Clark told him that he hadn't seen any. He finally found some evidence of our leasing but by that time it was too late and we had taken practically everything we wanted. He did buy some leases, thereby thinking maybe he was breaking our block but all of them were outside my guidelines. I later came back, about six months later with practically the same crew and we took the block at Mallalieu, which is producing. . . . The block around Ruth turned out to be a salt dome. (Vaughey March 11, 1987)

Other Companies

Few other companies sustained serious exploration in the southeastern states. Sun kept its highly regarded geologist, Donald Monroe, resident in Mississippi, but the company mostly made checkerboard lease plays, following other companies. Ohio kept Jerry Walden in Amory for several years doing a small amount of drilling and later drilled the two wells at Mobile. Overall, however, Ohio was not a consistent player. The same is true for Ramsey Petroleum—after drilling a dry hole on the Jackson Fault (Alabama), the company "walked away."

Humble and the Texas Company mainly made checkerboard lease and mineral plays. Humble's resident geologist of seven years, Paul J. Fly, resigned and moved to Texas in 1937. Shell kept a staff in the area primarily to collect geological data.

CHAPTER 22

Other Significant Events Affecting the Oil Business, 1926–1938

1930

In 1930, the largest field yet to be discovered on the North American continent was found in eastern Texas. Drilling spread through five counties until the field covered 134,000 acres with more than 27,000 wells. The East Texas Field became a giant with some six billion barrels of reserves. Steffey visited the field in April 1931, reporting that a completed well at 3,600 feet could be turnkeyed for $16,000 and that oil was selling for 17 cents per barrel.

1933

A publication by the U.S. Bureau of Mines written by Paul Weaver stated:

> The increase of geophysical surveying during 1932, particularly in the Gulf Coast region of Texas and Louisiana, is noticed. Gravity and reflection seismograph surveys were used almost exclusively. . . . The total outlay for gravity and seismograph surveys in the Gulf Coast region is now approximately $175,000 per month.

A 1933 security law was passed by the Congress prohibiting the use of the U.S. mail to raise funds. As raising money through the mail to drill wells was an old custom in oil circles, many prominent oil people who were unaware of the implications of this law continued to operate as before.

The 1933 Security Act resulted in a witch hunt. "Did you or did you not offer interest in an oil deal through the mail," oil independents were asked. One letter was enough to convict. Sentence: one year in prison. Soon many prominent independents were being charged with fraud. Within the next three to four years, many were convicted and sent to prison, at including least one from Mississippi.

On March 15, 1933, Steffey reported:

> Owing to our funds being tied up in one of the state banks here, we have been unable to leave town. Therefore, we have not been able to reach the folks operating to learn exactly what progress has been made since last report. With the opening of some of the banks beginning Tuesday, March 14, we expect operations will be resumed.

1935

The U.S. Coast Geodetic Survey began a project to set bench marks in Mississippi and Alabama.

The Public Utility Holding Company Act of 1935 was passed, requiring utility companies to dispose of all but one utility operation.

United Gas announced that it had bought the Youree property situated away from the business section of Shreveport for the sum of $75,000 and would begin building a new headquarters building immediately. The company's headquarters would be moved from Houston when the building was completed. The Rodessa discovery by United and its increased operations east of the Mississippi River prompted this move.

1938

The Natural Gas Act became effective June 1, 1938, making gas companies subject to regulation by the Federal Power Commission.

Technological Advancement

Technical developments in the oil business had been dynamic between 1926 and 1938. Drilling depths below 4,000 feet were difficult in 1926, but depths below 8,000 feet were commonplace by 1938.

It was possible to cement casing by pumping down a rubber plug by 1938. This forced all of the cement out of the pipe and eliminated the necessity of "drilling the plug."

Perforating with bullets allowed a productive zone to be opened into the cased well bore with a precision that was not possible with open hole completions.

The electric log pinpointed depth intervals of oil and gas sands that took out most of the guesswork. Previously, it had been necessary to rely on depths estimated from shows in cutting or cores.

Reflection seismograph made it possible to map structures at depths thousands of feet below the surface of the ground.

Mississippi, Alabama, and Florida were poised to be the first states where virtually every producing well and dry hole would be surveyed with an electric log. Exceptions were those wells at Amory, the Jackson Gas Field, and wildcats drilled before 1934—these represent a relatively few shallow wells compared to the drilling that was to follow.

Part III

The 1939–1941 Era

CHAPTER 23

Mississippi, 1939–1941

The golden age of petroleum in Mississippi was the decade of the 1940s. Over half the oil and gas ever to be discovered in the state was found during this period. When the boom that oil men dreamed of arrived, Gulf and United Gas, having paid their dues with years of persistence, were to be the big winners.

Nine years had passed since the discovery of the Jackson Gas Field. During these nine years, scores of wildcat wells had probed almost every portion of Mississippi with no success.

Each dry hole represented a substantial financial loss to those who invested in it. The vision that prompts oil men to rationalize many consecutive failures and still believe petroleum can be found eludes the reasoning of most other people. In the late 1930s, an air of optimism appeared to be growing in Mississippi. New faces appeared, and geophysical activity was on the rise.

Mississippi Offshore

The first offshore exploration was started. Phillips Petroleum Company leased all of the Mississippi Sound, some 700,000 acres, from the state.

Beginning May 3, 1939, Phillips commenced shooting the vast water-covered area utilizing Seismograph Service Corporation equipment. A crew of 25 men was necessary to operate the boats, shothole drilling, and seismic equipment. It was a pioneering undertaking. Some 236 miles of shooting had been completed by the end of the year, but no structures were found.

The company continued seismic work into the first months of 1940, but abandoned the project around mid-year. This had become an incredibly expensive venture.

Gulf leased Cat Island in the Mississippi Sound and undertook geophysical work about the same time.

The Tatums

Needing gas for its pipeline, Willmut Gas and Oil was involved in a number of wells. One that made honorable mention in one of Steffey's report of June 1939 was near Clinton, Mississippi. The well burned a small flare, but could not be made into a commercial producer. Steffey's comments on this well:

> The Tatums of Hattiesburg . . . as much an institution in this "neck of the woods" as the Roosevelts of Hyde Park . . . are trying to bring in a gas well on their location known as the Cleve Love, S. F. Johnson No. 1, near Clinton, Mississippi (Hinds County). This well has had as much care and nursing as a month old baby these past few days.

Willmut also acquired the old Amory Petroleum Company block at the bankrupt sale. Afterward, the company purchased many new leases at Amory and drilled another well, which was unsuccessful. Willmut had some luck, however, on a number of wells in Illinois. The company had its own drilling rigs by 1940.

Sun Oil Company made a deal with W. S. F. Tatum to run a magnetometer survey on his large land holdings in southern Mississippi. With the information gathered by Sun's work, the Tatum Lumber Company drilled a well on its own lands to discover the Tatum Salt Dome in October 1940. The land over the dome was then leased to Freeport-Sulphur Company for sulphur rights only, the Tatum Lumber Company retaining oil and gas rights. In 1941 Freeport-Sulphur began drilling a series of tests into the cap rock.

Freeport's second well actually found an "over-hang" of the salt. Tatum exercised a right to take the well over and drill it deeper, but with no success.

Jackson Drills its Own Gas Wells

By 1939 only 40 gas wells remained of the original 137 that once produced in the Jackson Gas Field. The city of Jackson was destined to be short of gas during the winter of 1940. United Gas was working on a contingency plan to build a pipeline to bring gas from the Monroe Gas Field should new supplies not be found in Mississippi.

Much ill will developed toward United Gas for selling off the "city's gas," through its pipelines to other customers in south Mississippi, Louisiana, Alabama, and Florida. Jackson was still enjoying the very low-cost gas (30 cents per MCF).

United Gas informed the City Council:

> We are going to lay a pipeline from Delhi [Louisiana] to supply Jackson. Of course the price of gas was going to go up. . . . So the City Fathers got properly irate and said 'you can't do this to us. There must be more gas out there,' and they hired Dad [Geoffrey Jeffreys] to come back and tell them what to do. . . . So the City decided they would drill five wells in the City limits which would be City owned and which would prolong the supply of local gas. (Jeffreys, 1989)

Actually, six wells were drilled and all were dry with the exception of one at the intersection of Grayson and Adelle Street in Belhaven. This well was completed in September 1940 for 30 million cubic feet of gas per day.

The City Councils decision to drill these wells was doomed to failure from the outset. Over the years as gas was taken out of the gas rock, the water table in the reservoir rose to occupy the space previously bearing gas. Drainage was very uniform, and locations that would have been good in the early stage of the field had since "watered out." Any new wells would only share the rapidly diminishing amount of gas remaining in the reservoir with the existing wells.

Jeffreys warned the city not to pull the casings from these dry wells as there were "charged," shallow sands that might blow out. His warning was ignored, however, and in early 1941 the City of Jackson Fee No. 2 well blew out when the casing was being pulled. White sand coated the neighborhood in the vicinity of the well located at North State Street and Riverside Drive.

Young Geoff Jeffreys noted:

> I was coming home from Central [high school] one afternoon . . . and I heard a train, a freight train going through north Jackson . . . and I said What's this, a roaring freight train? Then the closer I got to Bailey [junior high school], the louder it became. I then broke into a run. . . . There behind the fire station this gas well was just blowing wild. . . . An old casing puller had gone to the City and gotten permission to pull the casing. . . .
>
> It cut the rig down like a sandblaster, just cut the metal bars of the rig. . . . It looked like butter as it melts. Sand was being spewed all down-

wind toward Euclid Street. There was a foot of sand on top of the nearest homes and about a block away it finally got down to two or three inches. (Jeffreys, 1989)

Fine white sand covered all the houses, trees, cars, roads, and lawns in the vicinity. It had the appearance of an early snow. (Lambdin, 1989)

The well bridged over and died on its own. No lawsuits were filed. It was later discovered that the one successful well drilled by the city was drilled on the wrong lot, not on city-owned land, due to a survey error. The city of Jackson soon got out of the gas business.

The city of Canton had been more fortunate with two gas wells at Belhaven and added a third gas well at the corner of Riverside Drive and Peachtree Street in 1941.

United's Search for Gas

United Gas, through its subsidiary, Union Producing Company, was making every effort to find new gas supplies in Mississippi. Efforts were redoubled in 1939 with supplies running short. In less than one year customers would have to be curtailed. Union Producing geologists combed the area around Jackson with seismic crews working full time.

Flora

A seismic structure had previously been found and leased near the town of Flora, just north of the Jackson Gas Field. Union Producing began drilling this prospect in March 1939.

E. B. McGehee purchased one-quarter of the minerals under 1,800 acres of the J. C. Anderson land on which this well was being drilled for $12.00 per acre. The price for minerals was going up. On open-hole test, the well recovered live oil with saltwater at a depth of 4,400 feet from the gas rock. Pipe was set, but the well was not commercial because of the large amount of saltwater. (Later drilling by other operators would prove up the Flora Field.)

Midnight Dome

From this location Union Producing Company moved its rig to a structure known as the "Midnight Dome" in Humphreys County. This was the prospect to which company geologists gave the best chance of having a new field. After two months of drilling, Union's C. B. Box No. 1 well was abandoned at 5,652 feet on August 24, 1939. The company had discovered another igneous intrusion similar to the Jackson Dome but smaller. Some gas rock-type reef was developed on the structure but was not productive.

The company had by then made a fateful decision to drill the structure at Tinsley. This would be the turning point for oil in the southeastern states.

CHAPTER 24

Tinsley Field

On April 12, 1939, a press release by the Mississippi State Geological Survey made public a structure that had been found by Fredrick F. Mellen some 37 miles northwest of Jackson:

> A structural "High" in Yazoo County has been discovered by Fred-eric F. Mellen, supervising geologist of the WPA-Mississippi State Geological Survey, in the minerals survey of the county, in which the Yazoo County Chamber of Commerce and Board of Supervisors are acting as co-sponsor.
>
> The first indication of the structure was noted in October, 1938, when a thin bed of bentonite in the Yazoo Clay member of the Jackson formation, a bed believed to be a reliable stratigraphic marker, was determined to lie at . . . approximately the same elevation of the bentonite bed at two places 9 1/2 miles apart. . . .
>
> In checking his Yazoo County stratigraphy in February, 1939, Mr. Mellen discovered 12 feet of the Moodys Branch marl member of the Jackson formation on Perry Creek, a mile southwest of Tinsley in an area where normally only younger Yazoo clays should have been exposed . . . show, therefore, the Tinsley structure to have a northward contour closure of at least 135 feet a structure so favorable for oil and gas accumulation as to warrant further geologic study and seismographic exploration.
>
> Although the existence of the Tinsley structure in a thickly loess-covered area is based largely on the evidence of a single outcrop of the Moodys Branch marl on Perry Creek, the structure of the higher bentonite bed tends to corroborate its presence, although seismologically surveyed,

and surrendered, some ten years ago by Amerada, nevertheless, the Tinsley structure should be further explored, especially with a seismograph, to determine whether or not the subsurface structure is sufficiently pronounced to warrant a commercial test well, and especially should it be further explored for the reason that it lies less than 35 miles northwest of the Jackson Gas Field.

Because of the great expense involved in oil and gas exploration and especially in deep drilling tests, this press notice is being released by William C. Morse, State Geologist, who spent a few days in the field, and by Frederic F. Mellen, the discoverer, only on the condition that this article be accepted in its entirety.

The faint surface indications mapped by Fred Mellen were clues that experienced geologists interpreted to mean that a much more pronounced structure might be present with depth. Successive layers of sediments deposited over the structure for millions of years had almost obliterated surface evidence of its presence.

The magnitude of the Tinsley structure would stagger the imagination of even the most optimistic geologist, had they known. A giant mound of salt some eight miles in length and three miles in width lay buried beneath the rugged hilly land of Yazoo County.

Growing at a rate of probably no more than a fraction of an inch per century, this huge salt mass had risen to a height of more than 10,000 feet above its original base. A broad faulted anticline was formed in the beds above the salt mound. If the salt core were brought to the surface, it would have formed a mountain with the relief of Pikes Peak. As it was, the peak of this salt mountain would be found at a depth of 11,000 feet (two miles) below the surface.

Union Producing Company read the press release and immediately sent a paleontologist to check Mellen's work. Coincidentally, the company had a seismic crew shooting nearby in Yazoo County, and it shifted the crew to the Tinsley structure:

[Ed] Ricketts said that he had a seismic crew working north and west of Jackson for United Gas. He said, that a geologist in charge was a Mr. John Ivy who was gung-ho on this new seismic system and worried about the Jackson Field's near depletion. They had mapped a little nub (Flora) and had run a line out Highway 49, through Yazoo City and into the delta. He didn't remember the spring day date Mr. Ivy called but told him he had read a WPA paper about a marl outcrop near Tinsley Station and wanted him to tie it into the line they had run out Highway 49. So he took off some five miles east of Yazoo City and ran across to Tinsley. Greatly to his surprise, the only good reflector was hundreds of feet higher than the point they left up dip. They went on to map the structure. (Phillips, 1989)

Tinsley is located in the loess hills west of the Mississippi River floodplain. These huge prehistoric sand dunes are probably a result of debris being carried in the runoff after the last North American ice sheet began melting some 20,000 years ago. Over the following 10,000 years, a strong prevailing westerly wind built the dunes, which are now heavily forested to form high hills with deep V-shaped valleys. "In order to obtain satisfactory seismic reflection, it was necessary to shoot in the valleys," (Storey, 1990).

Within two weeks of the press release Union Producing had landmen buying leases on the structure at 50 cents per acre, for 10-year terms. The first lease taken was from Green C. Woodruff on April 28, 1939. The discovery well would be drilled on this lease.

Union Producing had some competition. Henry Toler, chief geologist for Southern Natural Gas, acquired a number of leases for his company and actually began buying two days ahead of Union Producing (National Oil Scouts, 1940). Toler told his wife, Ruth, "if the Tinsley structure does not produce, then we will leave Mississippi." Southern Natural Gas became one of the largest lease holders on the structure, second only to Union Producing Company.

On the recommendation of Bill Spooner, a Shreveport geologist, Jones-O'Brien of Shreveport acquired a 1,160-acre lease from Jennie Stevens at 50 cents per acre. However, most of this lease was assigned to Union Producing, after N. C. McGowen made a deal for Union to drill the lease in exchange for all the acreage except three 80-acre tracts to be held by Jones-O'Brien. C. W. Sharp of Shreveport bought half the minerals under the Stevens lease at $2.00 per acre. George Hunt and L. E. Ridgway of Jackson were also buying leases and minerals. T. H. Dinkins and J. C. Palmer were among those from Yazoo City buying mineral rights.

The WPA supplied money to Mellen to core-drill the structure, even as Union producing rushed to complete its seismic survey. Preston Furgus, district geologist for Union Producing at Monroe, checked the borings from the core holes and confirmed Mellen's conclusions (Storey, 1990).

Union's geologists were soon satisfied with the structure they had mapped and management made the decision to drill. No rigs were available in Mississippi, and one had to be trucked in from Louisiana. Normally the rig would have come by rail, but just a few months earlier, the Yazoo and Mississippi Valley Railroad had removed the rail siding at Tinsley, and there was no place to unload. After crossing the Vicksburg bridge, the trucks carrying the rig started up Highway 3 toward Tinsley, only to be arrested by the Highway Patrol for illegal overloading. The fine was $500.00, quite a large sum in those days. This forced the truckers to take a roundabout way through Jackson to get to Tinsley (Mellen, 1987).

It had been an incredibly short period of time by "oil field standards" between the discovery of the Tinsley Dome and the arrival of the Union Producing Company rig. Bulldozers and mules cleared a location on the Woodruff lease. The well was spudded on July 18, 1939. Preston Fergus was assigned to "sit" on the well, as J. B. Storey, Union Producing's resident geologist in Mississippi, was tied up on the Midnight Dome well.

On August 29, 1939, the well hit oil. This was what every oil man had been hoping for, an oil field in Mississippi. The Union Producing Company's G. C. Woodruff No. 1 well was officially completed on September 5, 1939, from open hole at a depth of 4,538 to 4,560 feet, flowing 235 barrels per day of 34.8 degree gravity oil.

The Steffey Report of September 2, 1939, recites as follows:

> Excitement prevails throughout the state and especially in and near Yazoo City and here in Jackson. Streets and Hotels alike are crowded to capacity and business is humming. The following is a resume of activity covering the "Tinsley block" or dome as it is familiarly known. Union Producing Co.'s-Green Woodruff Well No. 1 (Yazoo County, Miss.) on which a test was made at 7 A.M. August 29, 1939 as follows: Drill stem test made; fluid rose 4000' in 4" drill stem in fifteen minutes; 1/2" choke, top and bottom; 32.8 gravity oil; no water; when breaking off fourbles coming out of hole, oil blew through double-board; test taken from 4540' to 4560'; 4560' bottom of hole; Schlumberger log showed from 25 to 30 ohms and 115 millivolts; NOTE: Cemented 4538' of 4" O. D. casing; used 200 sax cmt. (Haliburton) August 31, 1939; plan to drill plug Monday September 4, 1939 and run tubing and wash well in Tuesday Morning. . . .

> Naturally the above statement when released to the general public was greeted with incredulity and considerable caution, as hereto fore the mention of oil east of the Mississippi River *had been considered possible but not probable,* despite the fact that company and independents alike had invested men, time and money in an effort to bring in a producer (OIL).

The producing zone quickly became known as the "Woodruff Sand."

News of the oil discovery in Mississippi resounded through the industry. People poured in from almost every portion of the country. The rural roads in Yazoo City were crammed with cars, mostly leasebrokers hoping to find open leases near the discovery.

Union Producing had hoped for a large structure but had greatly underestimated its size. Even though the company held a sizable lease block, drilling would soon spread beyond the Union Producing acreage so that many leases bought by other operators would prove productive.

J. P. Evans in Tinsley

James P. Evans from Shreveport had been very active in Mississippi since he bought the original lease block at Amory in 1926. On hearing of the oil discovery at Tinsley, he jumped in his car with his son and drove to the area.

> I remember we were riding down the road on the south end of Tinsley . . . and saw a man out by his mailbox [my father asked], "your land leased?" He said "no." [My father asked] "what's your name?" And he said "Dazell Johnson." "I am J. P. Evans. I bought a lease from you once before [Evans purchased the block for Amerada in 1930]. Do you want to sell me another one?" He said, "yes sir." Dad bought the Dazell Johnson lease, 145 acres. I think he paid $1.00 an acre for it and subsequently got six wells on it. (Evans, 1988)

J. P. (Jim) Evans, Jr., moved to Jackson in January 1940 in charge of the Evans oil interests and became a permanent resident.

The Confirmation Wells

Without the normal wait to see if the well would hold up, Union Producing immediately made two new well locations. The first was three-fourths of a mile northeast of the discovery well on the Jennie Stevens tract. The rig from the Midnight Dome was moved to this location.

Jennie Stevens's husband had acquired the 1,160 acres of land at Tinsley in 1927 when his delta farmland was flooded. He bought the tract to have a place to run his cattle until the Mississippi River flood subsided.

Union made the second new location on the W. B. Perry estate half a mile south of the Woodruff well. The Perry well was spudded September 23 and the Jennie Stevens No. 1 started drilling on October 6. Union Producing began laying a three-inch line from the Woodruff No. 1 to the Tinsley

Switch on the railroad. Events were moving incredibly fast. The Yazoo and Mississippi Valley Railroad rushed to put back the rail siding that had been taken up a few months earlier. On September 28, 1939, four railroad tank-cars containing 2,000 barrels each were shipped to the Standard Oil Company refinery in Baton Rouge. This was the first crude shipped from the field.

The second operator to commence drilling at Tinsley, was Jones-O'Brien, Inc., on one of its 80-acre tracts "carved out" of the Jennie Stevens lease. This well was spudded on October 6, 1939.

Union Producing experienced a terrific letdown when the Woodruff Sand was not found at 4,400 feet in the Jennie Stevens No. 1, the confirmation well. The company's geologists were all smiles a few days later, however, when the well found oil in the Eutaw Sand at 4,800 feet. The well was completed for 250 barrels per day of 34.4 degrees gravity oil. The zone became known as the "Stevens Sand."

As more wells were drilled, it eventually became apparent that the Woodruff Sand was actually the northwestern edge of the Jackson Gas Field gas-rock reef. As the Woodruff Sand is traced southeastward, it thickens rapidly and assumes a reef facies that reaches some 2,000 feet in thickness around the periphery of the old Jackson structure. Union's first Jennie Stevens well had been located beyond the feather edge of the reef.

The Union Producing Company Perry No. 1 found oil in the Woodruff as did the Jones-O'Brien No. 1 Jennie Stevens. With four wells producing in the field, activity accelerated. By the end of 1939 Union Producing Company and Jones-O'Brien were drilling 15 wells in the fields.

The beginning of 1940 found Union Producing Company producing some 1,800 barrels of oil per day from six completed wells. Some 114,972 barrels had been produced during the four months since the field was discovered. By February, two additional operators, Hassie Hunt and E. C. Johnston, both established producing leases in the field.

Southern Natural Farms Out

In late 1939 the Steffey Report announced that Southern Natural Gas had made a location for a well on one of its leases at Tinsley. Then occurred a decision that greatly changed the fortunes of many companies and individuals. Southern Natural Gas decided not to drill Tinsley, but to "farmout." In hindsight, this seems unbelievable as there were already several producing wells adjacent to their leases.

Bill and Emmett Vaughey, a new pair of independents in Mississippi, learned of the decision. Going to Birmingham, the Vaugheys were granted a 30-day option on all of Southern Natural's leases with the obligation to drill a minimum of four wells with Southern Natural retaining a one-eighth overriding royalty. Vaughey & Vaughey offered the deal to almost every company in the business, but no one believed a profit could be made on a lease burdened with such a royalty. Their option ran out.

Hassie Hunt, son of H. L. Hunt of Dallas, learned of the farmout and took the deal from Southern Natural Gas on similar terms offered to Vaughey & Vaughey:

"I did the dip work [seismic interpretation] for Hassie Hunt in the winter of 1939 or early 1940 on what we called North Tinsley" (Phillips, 1989).

When drilling in the field was completed almost every acre of the Southern Natural Gas farmout produced. Hassie Hunt would eventually become the second largest operator in the Tinsley Field with some 50 oil wells. Had Southern Natural Gas drilled its own leases, it would have greatly increased, possibly even doubled the value of the company.

Slick-Urschel

Soon another operator began drilling on the Tinsley structure on the Tom B. Slick fee land. Slick had purchased the land in 1916 after seeing it from a train window and immediately having a premonition that it was underlain by oil. The plantation was now located adjacent to production.

After Slick died in 1930, his comely widow, Bernice, married his business manager, Charles F. Urschel. Urschel attracted nationwide attention in 1933 when he was kidnapped and held for ransom by Machine Gun Kelly and Kelly's wife from Mississippi, the former Cleo Coleman. The ransom was paid and Urschel released.

The land was owned by the Slick estate at the time of the Tinsley discovery. The trustee of the Slick estate leased 825 acres of the tract to Slick-Urschel Oil Company; the former Mrs. Slick and her children shared a three-eighths royalty. Eventually 30 oil wells would produce on the property. Tom Slick had been right again. Not bad for a hunch made from a train window.

The Field Expands

By mid-May 1940 there were 49 producers in the field flowing 7,350 barrels of oil a day with an allowable of 150 barrels of oil per day per well (the allowable had been 300 barrels of oil per day prior to April 1940). The oil was being transported by tankcars to the Standard Oil of Louisiana Refinery in Baton Rouge. Steffey reported that Union Producing managers:

> have procured fifteen lots in "Tinsley," Yazoo County for the erection of an 80,000 barrels storage tank, supposedly the largest in the history of the Oil industry. Also, Union is now erecting housing on property near Tinsley for employees.

Union Producing Company stationed B. G. Barnes as field superintendent and J. L. New as production superintendent in the Tinsley Field, early in the year.

At the end of 1940 a total of 133 oil wells were in production in the field, having yielded 4,210,022 barrels of oil during the year.

Even though Union Producing was the largest operator, Hassie Hunt, Sohio, Slick-Urschel, Magnolia, and Jones-O'Brien were among the other larger operators. In addition, other parties that would drill producing wells in the field were Sinclair, Magnolia, Plains, Love Petroleum Company, James P. Evans, Ginther & Warren, Lyons and Prentiss, Midstates, Amerada, Frankel, Hodges & Culberson, and Beckett.

The Spacing Controversy

As a fledging oil state, the course of events at Tinsley would set the precedence for regulations that would govern the drilling and production of Mississippi's petroleum resources. Soon disputes between oil operators were aired before the the state Oil and Gas Board. This three-man board was chaired by the governor and included the attorney general and state land commissioner.

The governor was Paul B. Johnson after his inauguration January 15, 1940. One of the Oil and Gas Board's first actions at Tinsley was to establish 40-acre spacing for the Field (only one well per 40 acres). The rule was adhered to during 1940, with the first 150 wells drilled on 40-acre spacing.

Several operators, however, objected to the spacing. R. J. Whelan of Marshall, Texas, asked for permission to drill three wells on the railroad right-of-way as a special exception to the spacing rule.

In February 1941 the board granted a permit waiving the 40-acre spacing rule by allowing a permit to drill on a strip of land only 125-feet-by-38-feet (less than one acre). This would cause a flood of unnecessary drilling during the development of the field.

In April, 10-acre spacing was granted to two tracts, Gholsons and Lutz

Johnson, despite substantial opposition from other operators. Buzz Morgan reports in his Dixie Geological Service:

> Since the Gholsons' and the Lutz Johnson applications have been granted, there has been somewhat of a deluge of applications for permits to drill on tracts of less than 40 acres. (Dixie, May 8, 1941)
>
> If these permits are granted and the wells are drilled, it is reasonable to expect the 40 acre spacing rule to be abrogated. . . .
>
>   F. H. Broocks' application to drill a well on a 1 3/4 acre tract was granted! However, the Union Producing Company appealed to the Oil and Gas Board, which will meet on Tuesday, May 27th, in Governor Johnson's office and decide this case. (Dixie, May 22, 1941)
>
> The F. H. Broocks-Union Producing Company controversy is interesting almost to the point of "smelling"! (Dixie, May 29, 1941) [the governor's former law firm handled the case.]

Once the door was open, Hassie Hunt became probably the biggest violator of the spacing rule and drilled more exceptions than any other operator. Hunt also put a large pump on one of his wells capable of pulling a thousand barrels per day. Union Producing Company stayed steadfastly with 40-acre spacing despite drainage the company may have suffered from other company wells in a common reservoir:

> Section 31 now has 28 wells producing, 2 wells drilling, and 2 locations. There is a possibility that 4 more wells will have to be drilled. On normal 40 acre spacing this section would have only 16 wells. The average spacing will probably end up very near one well to 20 acres. This will not be as dense as the southwest end of the field, which is starting on a 10 acre spacing. (Dixie, September 4, 1941)

In view of the inequity being created by allowing drilling on small spacing, the state decided to establish production allowables based on the size of the unit. A unit of less than 40 acres would receive a proportionately lower allowable.

> The State Oil and Gas Board met in the Senate Chamber at 10 o'clock a.m. on Monday, August 4th, [1941], to hear proposal for establishing proration for the Tinsley Field. As soon as the meeting was called to order, J. B. Cox, who was representing E. C. Johnston, proposed that the hearing be postponed until after the operators could hold a meeting and discuss the situation. Governor Johnson ruled that the operators had been noti-

fied in advance, and that the hearing would continue as scheduled. He then asked to hear the proponents first. There was then about two minutes of complete silence, with a few snickers at the end. E. L. Brunini, J. P. Evans' representative then reiterated Mr. Cox's request for postponement of the hearing. No action was taken and the Board retired from the Chamber. . . .

The attitude of the operators as a whole seems to be definitely opposed to any form of proration, other than some sort of Gentlemen's (?) agreement. (Dixie, August 7, 1941)

Since the proration scare at Tinsley the per well withdrawals seems to be more equitably distributed than heretofore, but the operators are still under a form of pipe line proration. (Dixie, August 21, 1941)

Drilling continued in the Tinsley Field through 1941, with 322 producing wells by year end. Cumulative production for the year was 15,277,270 barrels, with Union Producing's share having reached 30,000 barrels per day. But, the over-drilling took its toll. Reservoir pressure dropped steadily, and before the end of 1941, every well in the field had ceased to flow and was "on pump."

Well Locations

Another problem was that some operators falsified reports of well locations:

The numerous location corrections given in this report are unavoidable. A few of the operators survey and stake their locations before applying for drilling permits, while others apply for their permits "within 150 feet etc.," and then locate their well according to the topography, and in quite a few cases toward the higher contours. In several instances at Tinsley some of the operators have had to hire a surveyor to find out just where their neighbor's well had been "fudged" to. Several of these corrections have taken some peculiar wiggles out of the contour maps. This annoying condition can be eliminated if the Oil and Gas Board will require that all locations be surveyed by a Certified Engineer, and that the correct location be sworn to by the engineer, and filed with Oil and Gas Board within a certain number of days after permit is applied for. (Dixie, June 5, 1941)

Transporting Tinsley Crude

It is unusual that a Field the size of Tinsley, which is now producing over three quarters of a million barrels per month, does not have a pipe line outlet. However, the market demand for Tinsley crude is scattered almost to the four winds, and the Standard of Louisiana at Baton Rouge is the only purchaser close enough to justify laying a pipe line, and the freight on their shipments is partly paid by the producer. (Dixie, July 24, 1941)

The Tinsley Field production has increased during the first six months of 1941, until it is now in 25th place for the entire country. (Dixie, August 21, 1941)

An attempt was made to construct a pipeline:

Union Producing Company planned to build a pipeline to Vicksburg and had surveyed, taken options, etc. . . .
The railroad stopped the line by putting on a fantastic unusual rate. They eliminated all overhead, capital, etc., and simply fixed a rate to Baton Rouge covering only the actual expenses of running train and tank cars. (Brunini, 1990)

The rate of 3.5 cents per hundred pounds of freight in 1942 was very low considering that the oil was transported a distance of 207 miles and the Yazoo and Mississippi Valley Railroad shared the fee with the Illinois Central Railroad.

The Vicksburg Chamber of Commerce, the Johnson Refinery, and several oil operators contested the rate before the Mississippi Service Commission, but to no avail. Not until after the war would a Tinsley pipeline be constructed (Brunini, 1990).

Tinsley Crude Prices

On April 24, 1941, Dixie Geological Service reported,

Tinsley crude was being purchased on a base price of 96 cents per barrel less approximately 5 1/2 cents per freight equalization and 6 cents gathering charge. This netted the operator approximately 84 1/2 cents per barrel.

The following week, Buzz Morgan noted,

also to be deducted is 2 cents for taxes, 15 cents average lifting cost, less the royalty which is generally 1/8th or 1/4 as was the case with the Southern Natural farmout. The operator would therefore be netting approximately 57 cents with the 1/8th royalty or 47 cents with the 1/4 royalty.

Further reports indicate that on May 22nd the price was raised by 10 cents per barrel.

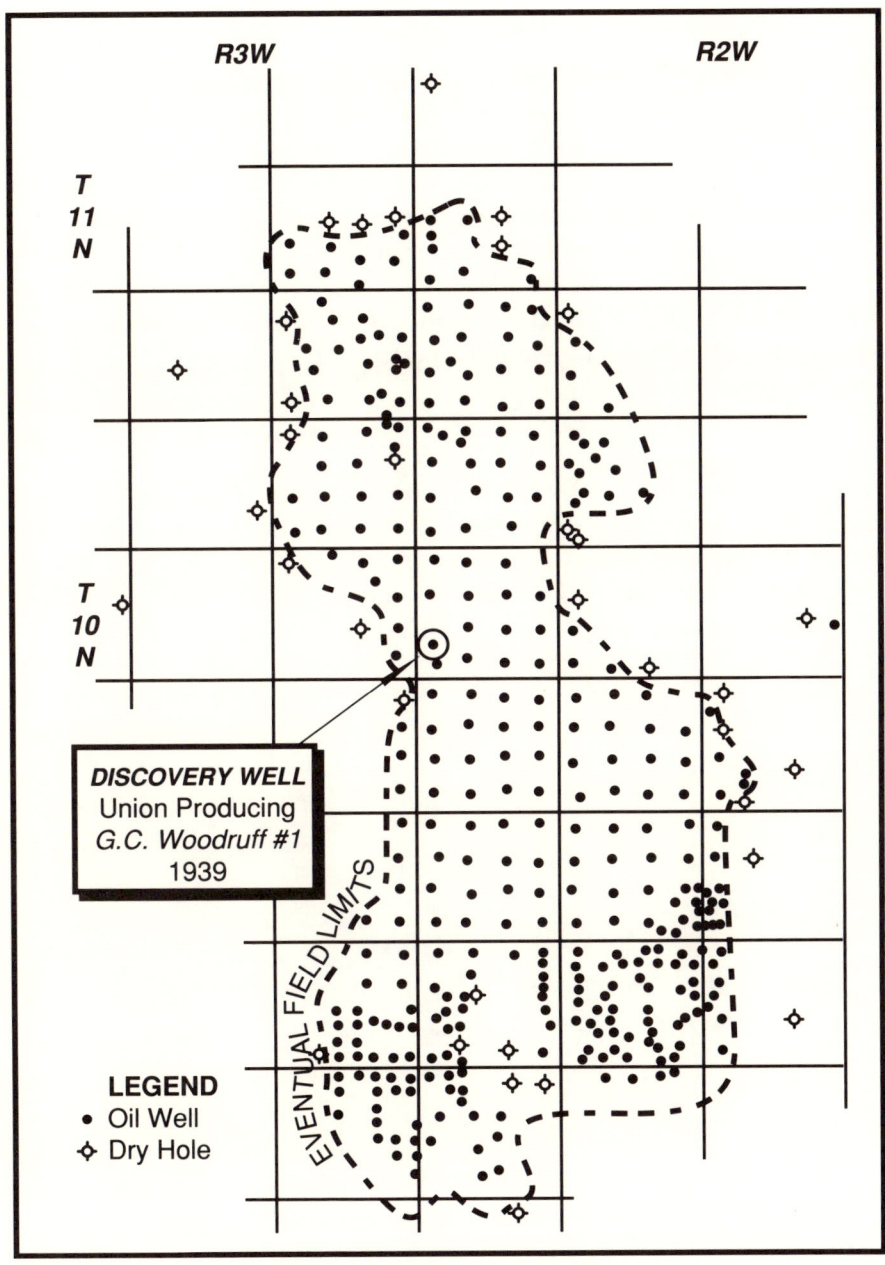

FIG. 7 Tinsley Oil Field, Yazoo County, 1945, Mississippi's first commercial
oil discovery

New Refineries

The availability of Tinsley crude in Mississippi gave rise to several new refineries, creating competition for the Standard of Louisiana Refinery at Baton Rouge. The first of these was built near Yazoo City by J. Constantin of Paris, Texas, under a charter granted by the state for the Paluxy Asphalt Company, headquartered in Talco, Texas. Construction began on May 1, 1941.

The second refinery sparked by the Tinsley discovery was the Delta Refinery at Memphis, Tennessee. Crude was to be delivered to a river terminal by barge. Construction was also to begin in May 1941. The same year, the E. C. Johnston refinery was put into operation at Vicksburg, recovering primarily asphalt.

Tinsley's Importance

The discovery of the Tinsley Oil Field in 1939 is considered in oil circles to be the most significant event to have occurred in the history of petroleum in the southeastern states. The Mississippi River barrier had been broken once and for all. Through December 31, 1988, Tinsley Field had produced a total of 219,494,277 barrels of oil, Mississippi's second largest field.

CHAPTER 25

The Dixie Geological Service

In 1938, Robert Steffey checked into the Veterans Hospital in Gulfport to recover from a "social disease" (this being before the days of penicillin). For the next three years, Fred Edam ran the Steffey Report. In March 1941 Steffey returned to Jackson to announce that he would take over the reporting. He also planned to move his headquarters from the Edwards Hotel to the W. P. Bridges Building. For the next 30 days, the Steffey report took on the character of the old days, with a great deal of folksy news.

Buzz Morgan, one of the geologists from Gulf's 1929 surface-mapping crews, made a deal to buy out Steffey. Morgan started a new scout report, the "Dixie Geological Service." After his first report was published April 24, 1941, the Steffey Reports were discontinued. Those who knew Steffey at this time say that he drank heavily and apparently spent his last years at Whitfield Insane Asylum.

Dixie Geological became a first-class commercial scout report that faithfully reported a complete resumé of all wells drilled in the southeastern states for the next three decades.

Buzz Morgan enticed his old friend, Bud Norman, to return to Mississippi to work with him. The December 11, 1941 scout report says:

WORDS CANNOT EXPRESS THE PLEASURE IT IS TO ANNOUNCE THAT MARVIN E. "BUD" NORMAN, OF FORT WORTH, TEXAS, WILL BE ASSOCIATED WITH ME IN THE OPERATION OF THE DIXIE GEOLOGICAL SERVICE.

CHAPTER 26

Two More Discoveries in Mississippi

Seven months after Tinsley was discovered, a well was spudded that would tap the second oil field in the state. This new drillsite was thirty miles northeast of Tinsley.

Kingwood Oil Company from Okmulgee, Oklahoma, assembled a block of 5,000 acres in Yazoo County near the little town of Pickens. A half interest in the block was sold to Exchange Oil Company, a subsidiary of Sinclair Oil Company, for $32,000.

The Kingwood Oil Company No. 1 Wilburn well was spudded March 9, 1940, and was completed on March 29, pumping at a rate of 240 barrels of 40.1 degrees gravity oil per day. Completion was from open hole at 4,878 to 4,889 feet. Production came from a Eutaw Sand that became known as the "Wilburn Sand." The Pickens discovery did not seem to be very exciting, since the well was making some saltwater with the oil. The well would not flow.

Six wells were drilled during 1940 in close proximity to each other, two by Kingwood and four by Exchange. The No. 5 Wilburn well was lost when the derrick pulled in during a drillstem test. After the No. 6 Wilburn well was dry in September 1940, the field appeared doubtfully commercial. No wells were drilled in the Pickens Field for the next year.

Sinclair had taken over operation of all the Pickens wells near the end of 1940. Cumulative production for that year was 285,000 barrels. Large amounts of water plagued the operations, and little significance was placed on the field's potential. Sinclair's top executive had an obsession for drilling on 10-acre spacing, which prevented the company from stepping out far

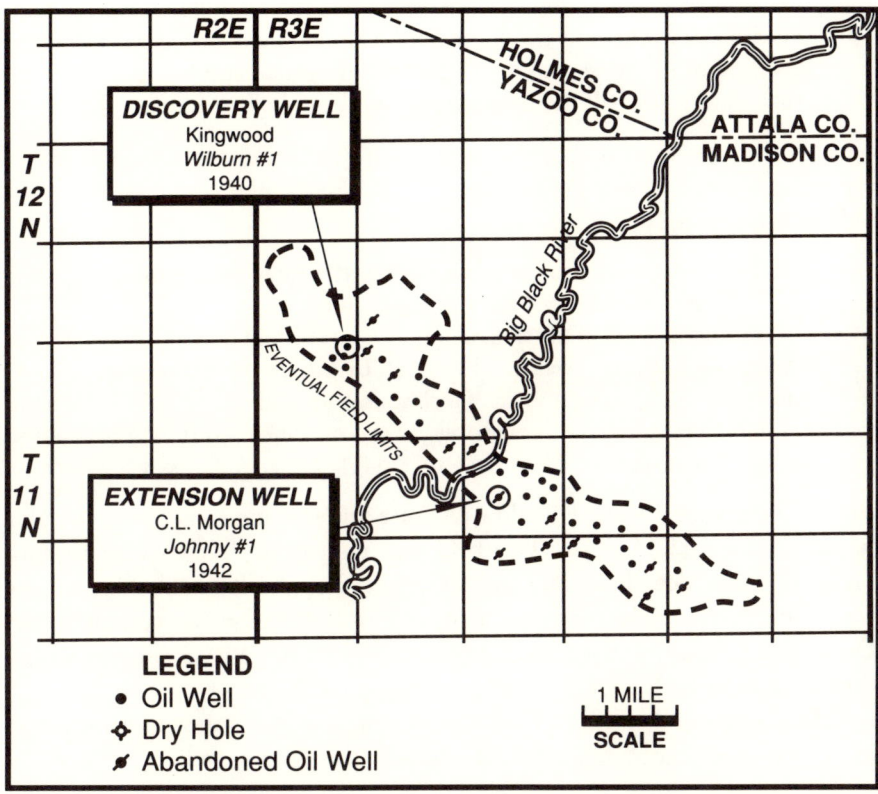

FIG. 8   Pickens Field, 1945, Mississippi's second oil field

enough to prove extensions to the producing area. By year's end, only three wells were commercially producing. During 1941 production was 218,000 barrels.

Not all companies were convinced that Pickens was a small field. After conducting seismic studies in the area, Carter Oil Company had assembled a block of leases to the southeast. Between the Carter block and the Sinclair leases, Phillips had leased all available land. Buzz Morgan obtained a farm-out from Phillips Petroleum through the Jackson manager, M. E. (Dutch) Miesse, with the obligation to drill a well two miles southeast of Sinclair's wells. The No. 1 Johnny began drilling on October 23, 1941.

In November 1941 oil was tested from the Wilburn Sand that had produced at Pickens. Buzz's son, Dan Morgan, recalls:

> The only thing I can remember, the river was all over the bottom-it was flooded. Of course, I was just a little kid then. After they got through with

the drillstem test, and the logging, the water was about three feet deep down there in the swamp. They put me in a truck and had to winch the truck out because all the board roads had floated off. (Morgan, 1990)

Seven-inch pipe was set, but tubing was not available because of the shortage being created by war preparations.

Buzz Morgan recalls:

> When I was trying to get some tubing, the reply to one inquiry I made was: "What sort of interest do you have to offer us?"
>
> As soon as word of my difficulty got around I had five different offers of tubing, all from *independents*. Mr. J. P. Evans, Sr. of Shreveport, Louisiana, who has production at Tinsley and several other fields was kind enough to let me have enough tubing out of some used tubing he located in East Texas. (Dixie, December 12, 1941)

The well came in flowing with low pressure and making some water, similar to the Pickens Field wells. The field was named the Pickens-Sharpsburg Field with the idea that "Pickens" could be dropped if it did not connect with the Pickens Field. The No. 1 Johnny well was officially completed on December 5, 1941, flowing 162 barrels of oil per day plus 10 percent water, through a 8/64 inch choke; flowing tubing pressure was 250 psi.

Before year's end, Phillips Petroleum Company had staked four locations in the Pickens-Sharpsburg area, all offsets to the Johnny well.

Cary Field

The third oil discovery in Mississippi was made on a block taken by Vaughey & Vaughey in Sharkey County. Based on an old magnetometer survey, several thousand acres of the F. B. Houston estate were leased to the Vaugheys, who in turn sold the deal to the British-American Company.

The No. A-1 F. B. Houston Estate well was spudded in early April 1941, and in May pipe was set to test shows in the gas rock around 3,200 feet. After several months of testing, the well was finally put on pump and officially completed on July 31, 1941, at a rate of 35 barrels per day of 25 degree gravity oil. Another producer was drilled a few years later, but the field did not prove to be commercial. Total production was only 160,687 barrels before the field was abandoned:

> We almost discovered a [third] field in Mississippi. Some geologist who lived in Houston came to our Houston office and showed us a core drill structure that they had worked out near Cary, in Sharkey County. We

made a deal with them to drill the well and carry them for an interest. The land in that area was all owned by a family named Houston. They were nonresidents and had been lumbermen. We picked up several thousand acres from them and promoted the well. It came in making about 25 barrels a day; we really thought we had found a new field. However, the production declined rapidly and at 90 cents a barrel [it was not commercial]. (E. A. Vaughey, March 11, 1987)

CHAPTER 27

The Oil Fraternity Expands

With the discovery of oil at Tinsley, the city of Jackson, Mississippi, became the undisputed headquarters for the oil and gas business in the southeastern states. Oil people poured into Mississippi.

The Companies

Union Producing Company moved J. B. Storey from Hattiesburg to Jackson with a promotion to district geologist in 1940. John Rogers was district manager and George Gilbert, landman. In 1941 Wilbur Knight came to Jackson as a geological scout for Union Producing.

Triangle Drilling Company opened an office in the Capitol National Bank Building; Eureka Petroleum Company had offices in the Kennington Annex; Minerva Oil Company of St. Louis and Lion Oil Company settled into the Tower Building (now Standard Life Building); Magnolia Petroleum Company, the Millsaps Building. Other companies moving to Jackson were Plymouth Oil Company of Sinton, Texas; Independent Exploration Company; and Sinclair-Prairie Oil Company. National Supply Company of Fort Worth, Texas, opened an office in Yazoo City.

Two mapmakers, Globe Map Company of Shreveport and Edgar Tobin Aerial Surveys of San Antonio, came to Jackson. Tobin opened an office in the Edwards Hotel, with Fred L. Wohn in charge.

By the end of 1939, Leroy Francis, scout with Carter Oil Company, was stationed in Jackson and destined to become a lifetime oil scout in the area. Amerada Petroleum returned to Mississippi with Bill Sinclair stationed in Jackson. Union Sulphur Company transferred in B. G. Weeks, landman, and R. Merrill Harris, geologist, with offices in the Tower Building. Merrill Harris would become a prominent independent geologist in Mississippi.

Harry Sinclair, founder of Sinclair Oil Company, arrived in Jackson in February 1940 aboard his private railcar (with other company officials) to make a tour of Mississippi. During 1940 Atlantic Refining Company opened an office in Jackson with geologist Robert Spooner in charge and Frank Castleburry as scout.

By early 1940 Gulf, Stanolind, the California Company, Kingwood, Magnolia Petroleum, Carter Oil Company, Ohio Oil Company, Sinclair-Prairie, Barnsdall Oil Company, Fohs Oil Company, Humble, Shell, and Pure all had seismograph crews operating in the southeastern states. In mid-1941 The California Company moved its offices to New Orleans, from which the company controlled the Mississippi operations for the next three decades.

These were early arrivals in the army of oil people who would invade Mississippi. New companies opened offices in Mississippi almost daily. Most oil and gas exploration activities in Alabama and Florida were directed from Jackson. The oil community in Mississippi had grown to such an extent that several hundred key people were located in the city.

Independents

A flood of independents also descended on Mississippi. These included C. G. Koch, Ralph A. Johnson, J. B. Moncrief, W. J. and Sid Richardson, and I. P. LaRue.

Marshall Young of Fort Worth was soon active in the state with Roeser & Pendleton. W. O. Allen of Tulsa, Oklahoma, put together a 10,000-acre block in Warren County on the "Glass Dome" south of Vicksburg. Sid Wheless and Fred W. Shields from San Antonio became active in Mississippi.

Newcomers who would become some of the top independents of Mississippi arrived in 1939. E. A. (Emmett) Vaughey from Houston, of the Vaughey & Vaughey partnership, came first. Emmett had taken blocks in Mississippi for Standard of California in 1936 and 1937. Steffey reported Emmett's taking a block in Wilkinson County during June 1939, followed by 8,000 acres in Scott County in September and a similar block in Claiborne County by October.

His older brother, W. E. (Bill) Vaughey, arrived a few months later to lease a block of 30,000 acres in Clarke County, Alabama. Thus, the famous Vaughey & Vaughey team had arrived in Mississippi. The third member of the team was W. T. (Blackie) Blackburn, brother-in-law (Vaughey, 1987).

Bill Vaughey, the oldest, was born in Marseilles, Illinois, in 1904. His father, who had developed lung problems, elected to move west to drier climates. He moved the family to Tulsa, Oklahoma, which was then Indian Territory, where he became a pioneer in the early development of the oil business in the Tulsa area. Emmett A. Vaughey, Jr., was born there in 1908.

With his health still in question, E. A. Vaughey, Sr. moved his family further west to Albuquerque, New Mexico, where the children attended public school. Later both Bill and Emmett attended the University of New Mexico. Bill moved to Tulsa in 1925 to work in the maproom of Philmac Company, a forerunner of Phillips Petroleum Company, and soon was promoted to scout and landman. Following this, he accepted a position with Electric Bond and Share in Pennsylvania. In the 1930s, he moved to Texas to form W. M. Vaughey, Inc., an independent oil company. The new company drilled a series of wells in the Luling area of southeast Texas.

Emmett began his career in the oil business in 1927, also working as scout and landman for Philmac. Soon afterwards, he became an independent distributor of Phillips Petroleum products in Bartlesville, Oklahoma. In 1934 he sold this business and moved to Houston, where he became an oil lease broker.

Bill Vaughey sold his fledging oil business in 1937 and joined Emmett, where they formed the partnership of Vaughey & Vaughey. The team rapidly developed a very successful lease brokerage business and soon had a number of lease men working for them. William T. (Blackie) Blackburn, brother to Emmett's wife, Mary, became a junior partner in the late 1930s. Blackie was born in Bay City, Texas, in 1916.

Initially, Bill and Emmett were sent to Mississippi as lease brokers, buying for the California Company, Carter Oil, and Sinclair (Reese, 1986). Occasionally they bought leases for themselves. In 1944 they made the decision to become independents:

> When oil was discovered in Tinsley in 1939, we immediately received calls giving us the news. We came over here . . . [and] started buying leases for ourselves and also for major companies. The Carter Oil Company, a subsidiary of Standard Oil Company of New York, and the Sinclair Oil Company were two of our best customers and we mainly did all of their work. The other companies we worked for in Mississippi were Standard Oil Company of Ohio, Continental Oil Company, and several drilling contractors. . . .

In 1940, we had opened an office in Deposit Guaranty Bank Building and were running a crew out of there. We had brought two of the lease-men with us . . . who worked directly for us. We also hired and trained about three Mississippi people which comprise our leasing group. This would be in the years of 1940 and 1941. In 1942 we left Bill in Jackson, and I went back to Houston, Texas, to take care of our leasing business back there. In 1944 we decided to give up our brokerage business and confine ourselves to trying to find oil. This was almost necessary as you cannot do both. When a major company wants you to take a block, they call you into their office, show you the outlines of the structure and you have that very valuable information, you have to keep confidential and we always did. However, when you are trying to find oil for yourself, this sometimes comes in conflict, so we decided to abandon a very good bro-kerage business. . . . This was a very hard decision for us to make because we liked the brokerage business, it fed us a lot of information, paid well, and it is exciting. (E. A. Vaughey, March 11, 1987)

Don Reese

On September 1, 1941, Don Reese, a geologist with Sinclair, was transferred to Jackson from Houston, to begin a stay that would become permanent. Born in Lincoln, Nebraska, in 1905, Reese spent his school years in Cali-fornia, but returned to Nebraska to receive his geological degree from the university in 1927. He became Sinclair's East Texas district geologist during the development of the East Texas Field and later district geologist for the Texas Gulf Coast in Houston.

On moving to Mississippi, during the Tinsley boom, he replaced geologist Bob Gossett. Sinclair's landman in Jackson was Joe Snider. Reese recalled:

Who would ever think that you would spend most of your life in Missis-sippi? We didn't think so. I don't think anybody else did, and here we are. I spent most of my life in Mississippi, and I don't regret it. (Reese, 1986)

H. D. Easton

In Amory, geologist H. D. Easton and his son, H. D. Easton, Jr., both from Shreveport, went into partnership with P. J. McAlpine in 1939. Easton had been credited with discovering a number of fields in northern Louisiana, including Sugar Creek, Benson, Zwolle, and Converse. He had made the location for the discovery well at Amory in 1926. The new partnership drilled a series of wells in the Black Warrior Basin of Mississippi and Ala-bama over the next few years.

Bob Hearin

Robert M. (Bob) Hearin graduated from the University of Alabama in 1939 just in time to get in the oil play at Tinsley. A native of Alabama, born in Demopolis and reared in Montgomery, Hearin made one of the best decisions of his life when he headed for Mississippi with his new degree in hand. He drove from Tuscaloosa to Tinsley Field where he looked up an old friend, Tip Ray, a lawyer in Yazoo City. After working for Ray for a short time, he was employed by Union Producing Company as a landman and scout, and was involved in the development of the Tinsley Field from its early stages. Hearin would rise quickly in the ranks of United Gas and become a leading figure in southeastern oil circles during the 1940s and 1950s. His experience in the oil business was just the beginning of an extremely successful career that would make him one of the wealthiest individuals in the South.

Dr. Richard A. Priddy

As a result of the discovery at Pickens, Mississippi gained another resident who would become a well-known geologist in the state, Dr. Richard R. Priddy. Working as a geologist for Kingwood Oil (from Effingham, Illinois), Priddy was given credit for discovering the Pickens Field. He quit Kingwood to be appointed assistant geologist for the Mississippi Geological Survey succeeding Fred Mellen in March 1940. (Dr. W. C. Morris was director of the Survey.) Several years later, Dr. Priddy became dean of geology at Millsaps College.

Fred Mellen, Consultant

On March 11, 1940, Fredrick Mellen tendered his resignation with the Work Projects Administration and the Mississippi Geological Survey. Mellen had been engaged in this work for seven years, but decided to become a consulting geologist and open an office in West Point, Mississippi. He spent the next two years promoting several shallow wells in the Black Warrior Basin; all were dry.

While he was pioneering this area 10 years ahead of the gas play that was to come, all the action was taking place in southern Mississippi because of Tinsley. Mellen had received a great deal of notoriety from the Tinsley discovery. Unfortunately, his small financial resources and limited experience in the oil business resulted in no monetary reward from the monumental Tinsley boom. Had he elected to stay in Jackson and work for one of the companies to gain experience, he might have realized a much greater economic gain during his career.

Dave Harrell

One of the arrivals in Mississippi during 1939 who was destined to become a permanent resident was David Crozier Harrell. Born on April 19, 1904, Dave Harrell graduated from the University of Texas with a B. A. in geology in 1927. Two years later, he married one of his classmates, Olita Harrow, who had graduated with an M. A. degree in geology. She too had a distinguished career.

After a short stint with Roxanna Petroleum Corporation, Dave Harrell worked in Sumatra and Java for Standard of New Jersey from 1929 to 1932, and in Argentina from 1933 to 1936. From 1937 to 1939 he did surface geology in southwestern Arkansas, North and South Dakota, and northeastern Kansas for Carter Oil Company.

With the discovery of oil at Tinsley, Dave and Olita were transferred to Jackson, Mississippi, in 1939 and liked what they found. After years in the jungles of Asia, the wilds of Argentina, and the cold plains of the Dakotas and Kansas, Jackson looked very good. In 1943 the Harrells were threatened with another move, but chose instead to resign from Carter and accepted a job with Sun Oil Company as district geologist in Florida.

By strange circumstances, Don Monroe, Sun's district geologist in Mississippi, had moved to Hattiesburg because he would not share an office with Melvin Campbell, Sun's district landman in Jackson. Don Monroe was given an ultimatum by the Dallas office to return to Jackson or exchange jobs with Dave Harrell in Florida. He stubbornly elected to go to Florida where he remained for the rest of his career. By default Harrell was able to stay in Jackson and become Sun's district geologist for Mississippi. He would be president of the Mississippi Geological Society for the year 1942–43.

With the war having siphoned off many of the younger geologists, however, there were no other geologists to help him. He spent the war years doing everything himself, including a great deal of well sitting, often seven days a week. Olita worked for Sun as a paleontologist in Jackson until she retired. After 14 years, Dave would quit Sun in 1956 to become a consultant in Jackson for the remainder of his career.

Wil Knight

In 1941 Wilbur H. Knight was hired by Union Producing Company and transferred to Jackson, to assist J. B. Storey in the development of Tinsley.

Born in 1921, Wil Knight was the son of Samuel Howell Knight, chairman of the Department of Geology at the University of Wyoming. In 1942, Wil was inducted into the army, serving until late 1944 and returning to Union Producing Company in 1945.

In 1947, J. B. Storey was transferred to New Orleans as District Manager and Knight became district geologist of the Jackson district. He spent his entire career in Mississippi.

The Scouts 1939–1941

With the discovery of oil at Tinsley, the Mississippi Oil Scouts Association mushroomed. From ten members in 1938, the association expanded to 38 members by the close of 1939.

One of the newcomers was J. A. (Jimmy) Morgan, formerly of Modisett Drilling Company, who had become a landman for Smith County Oil Company (an Eastman Gardiner offshoot).

In 1940, Storey was president ("Bull Scout") of the Mississippi Oil Scouts. Members included Red Fidler of Tidewater, Bert Gamble of Lion Oil, Merrill Harris of Union Sulphur, and Brame Womack with E. B. Germany. By the close of 1941, the association was being reduced by the defense buildup, and membership dropped to 22.

The Mississippi Geological Society

The Mississippi Geological Society was formed in 1939. Membership was 101 members as of February 3, 1940. The first officers were Henry Toler, president, and Tom McGlothlin, secretary/treasurer. David C. Harrell became secretary/treasurer in late 1940. Companies represented were:

Arkansas Fuel Oil	Pure Oil
Amerada Petroleum Company	Quarter Oil Company
Byrd-Frost	Sells Petroleum Company
California Company	Shell Oil Company
Devonian Oil Company	Sinclair Oil Company
Exchange Oil Company	Southwood Oil Company
Gulf Refining Company	Stanland Oil and Gas
Kingwood Oil Company	Sun Oil Company
Magnolia Petroleum Company	Texas Pacific Coal and Oil
Mid States Oil Company	Company
Ohio Oil Company	Transwestern
Phillips Petroleum Company	Union Producing Company
Plymouth Oil Company	Woodley Petroleum Company

The geological society conducted a field trip for its members in northern Alabama December 6–8, 1940. The group stayed at the old Tutwiler Hotel at Birmingham where rates were $2.50 for a single room with bath or $2.00 each for a shared room with twin beds. The second night was at the Russel Erskine Hotel in Huntsville.

The society announced a dance for February 17 at $5.00 per couple. In 1941 McGlothlin became the second president and Harrell, vice president.

The New Orleans Geological Society was formed in late 1941:

> Their first regular meeting was held on Monday night, November [1941]. Urban B. Hughes, independent geologist of Jackson, Mississippi, gave a talk on the Geology of Mississippi. The attendance was about sixty and Mr. Hughes reports that there was a surprising amount of interest shown in Mississippi. (Dixie, November 6, 1941)

CHAPTER 28

Other Exploration Activities in Mississippi

After Tinsley, exploration activity throughout Mississippi picked up greatly. Sixteen wildcat wells were drilled in the state during 1939, along with the development wells at Tinsley. New drilling rigs arrived almost daily.

By January 1940 57 geophysical crews were working in the state, mostly doing seismograph. Thirty-five new companies were granted charters to do business in Mississippi, and interest steadily increased:

> At a recent convention of the American Association of Petroleum Geologists held in Houston, Texas, more interest was shown in Mississippi than any other area in the entire country. . . .
>
> The recent oil shows in the I. H. Morgan well in Stone County, Mississippi, has attracted attention to the southern part of the entire Gulf Coastal Area. (Dixie, April 24, 1941)

The Satartia Structure

Very close to the day that oil was encountered on the Tinsley structure, the Mississippi Geological Survey decided to release paper on a structure it had mapped in Yazoo County, the Satartia structure. Released on August 16, 1939, the memorandum for the press stated:

> Satartia, Yazoo County, lies fifteen miles south/southwest of Yazoo City and five miles southwest of the center of the Tinsley Dome. . . .
>
> In many places eastward and southward of Satartia, the hill slopes, gulleys, and the streams of the bluff region exposed the thin Bentonite beds of Upper Yazoo Clay. As early as November, 1938, members of the WPA-

Mississippi Geological Survey Clay and Mineral Survey of Yazoo County, suspected that the irregular elevations of the Bentonite outcrops were due largely to structural conditions. . . .

Still later, the finding of pieces of bitumen near the center of Section 6, Township 9N, Range 3W, seemed further to establish the basis for the belief that the Satartia region . . . (Steffey, August 19, 1939)

By September Carter had begun conducting seismograph over the Satartia feature.

Seismic Difficulties

In a June 1941 report Buzz Morgan mentions that "seismograph has been quite disappointing in Mississippi" and that the Sun Oil Company had resorted to large-scale core drilling. Other companies planned to do the same:

In certain areas in Mississippi, the surface beds acts as filters and very little energy is returned to the seismometers. If, in order to overcome this difficulty, it is necessary to employ methods used in other parts of the country, the cost will be prohibitive. (Dixie, June, 1941)

During 1941 Sinclair-Wyoming Oil Company was the leader in the search for oil in Mississippi, according to Morgan. Roy Halsey was Sinclair-Wyoming's division geophysicist in Jackson. "Sinclair completely covered Mississippi with gravity. Some of their structures were checked with seismic and unfortunately 'dug only to Tuscaloosa' " (Phillips, 1989).

Mississippi Drilling Statistics

By the close of 1941, nine salt domes had been discovered in Mississippi. In addition to Midway and Edwards, the new domes were Glass, Tatum, Doyt, Oakvale, Newman, Kings, and Halifax. It was discouraging that no production had been found on any of these. By the end of 1941, a total of 1,010 wells had been drilled in Mississippi, of which 425 were dry wildcats, 154 were gas wells, 332 were oil wells, and 100 were dry field wells.

New field discoveries included: one oil discovery in 1939 from (16) wildcat wells drilled; one oil discovery and five salt dome discoveries from 76 wildcat wells drilled in 1940; and two oil discoveries and two salt domes from 49 wildcats in 1941.

Gulf drilled the Newman No. 1 in Scott County, to 10,365 feet, setting a new depth record for the state.

CHAPTER 29

Alabama, 1939–1941

The search for oil and gas in Alabama was relatively subdued during the 1939–41 period. Beginning in 1939, the National Oil Scouts Yearbook carried a special section on Alabama. Six wildcat wells were drilled in 1939 followed by 14 in 1940 and 14 more in 1941. All were dry.

Leasing activity increased from 180,000 acres in 1939 to 2,300,000 acres by 1941. Humble was the biggest lease buyer with 1,300,000 acres. Geophysical activity greatly increased with Humble, Magnolia, Sinclair-Prairie, Texas Company, Atlantic, Stroke Petroleum, Superior, Shell, and Union Producing, all active.

Hatchetigbee Deep Tests

Probably the most significant development was a deep test by Union Producing Company on the Hatchetigbee Anticline. When Joe Modisett passed away in November 1939, he had assembled a block of leases at Bladen Springs around the dry hole he had drilled to 7,250 feet several years earlier. His intention was to reenter the well and deepen it with a major company partner. This would test the northwestern portion of the Hatchetigbee Anticline.

Union Producing Company took over the project after Modisett's death and reentered the well with the idea of drilling on to 10,000 feet. Problems were encountered with the old hole, and the well, the No. 1 McCorvey, was junked at 9,622 feet in June 1940. In spite of the problems, this set a new depth record for Alabama.

(in random order) Buzz Morgan, Bud Norman, Sparky McGlothlin, Henry Toler, B. G. Martin, Floyd Ayres, James Tierney, Jim Aimer, Victor Grace, Gail Montgomery, Felix Richardson, and Hasting Faulkner

Left: Bud Norman; *above:* J. T. (Sparky) McGlothlin.

The above three photos and following six photos are of Gulf surface mapping geologists at Meridian, Mississippi, 1929–1930.

McGlothlin (l) and Jimmie Tierney

Above: (l to r) Buzz Morgan, Sparky McGlothlin, and B. G. Martin; *left:* Buzz Morgan.

Jim Aimer (l) and Martha Mooney

Left: Martha Mooney (l) and Nell Cariker; Martha Mooney became Mrs. Buzz Morgan and Nell Cariker married Jim Aimer; *below:* Jim Aimer, unidentified map reader, and Tom McGlothlin.

Right: Ella Rawls Reader, 1903, during presentation to the Queen of England in London; *below:* stock sold by Ella Rawls Reader Stokely in her Mississippi Insane Asylum wildcat well, 1932.

The Love Brothers, 1912–1913, Caddo Lake, Louisiana. (l to r) Cleve Love, "Big Boy" Love, M. C. Chambless, and "Peanut" Love

Mendoza Well No. 1. (r to l) "Big Boy" Love, J. A. Roell, and "Peanut" Love

WHEN "MIGHTY MENDOZA" BLEW IN APRIL, 1930

Flowing gas from the Mendoza Well, April, 1930

Gas well blowing in on the Baptist orphanage just off Woodrow Wilson at Five Points (now Jackson Mall). After discharging the heavy mud it produced 35,000,000 cubic feet of gas per day, 1933.

Henry N. Toler, 1936

Dr. George C. Swearingen, Mississippi's first Oil and Gas Board Supervisor, 1932–1936

Above: Geoffrey Jeffreys, a prominent international geologist, who was active in Mississippi during the 1930s and 1940s; *right:* Dan G. Hughes and wife, Winnie, 1928.

Fred West (with briefcase), an active independent during the 1930s

The Hughes twins, 1936. (l to r) Dudley, June, Dan, and Jane

W.S.F. Tatum (Willie Scion Franklin Tatum), wealthy Hattiesburg lumber man and mayor, founder of Willmut Gas and Oil Company in the 1930s

Right: Fred Mellen, Mississippi's best known geologist, who discovered the Tinsley structure in 1939; *below:* the discovery well at Tinsley testing oil, August 29, 1939. The hole was drilled by Union Producing Company on G. C. Woodruff land.

Mississippi Oil Scouts, Jackson, Mississippi, 1939. (Top picture, front row, l to r) W. H. Cordell, Tidewater; J.O. Harper, Pure; Joe B. Wheeler, Stanolind; Maurice Miesse, Phillips. (Second row, l to r) W. M. Payne, California; Dan W. Cameron, Carter; R. S. Cate, Sun; Frank Castleberry, Atlantic Refining. (Third row, l to r) Joe Hemphill, Magnolia; R. M. Harris, Union Sulphur; J. H. McCready, Kingwood; C. E. Leary, Texas. (Fourth row, l to r) J. H. Bohlender, Humble; J. B. Storey, Union Producing; C. F. Kimball, Sinclair-Wyoming; R. C. Edwards, Gulf Refining. (Lower picture, front row, l to r) Clark Baker, Skelly; Gayle Griffing, Superior; Al Beasley, Shell; Henry Toler, Southern National. (Second row, l to r) C. W. Seedle, Arkansas Fuel; J. F. Michaels, Amerada; J. W. Eiserloh, Gulf.

Scouts visiting the Carter Oil No. 1 Abernathy, near Okolona, Mississippi. (l to r) J. F. Michael, Amerada; A. T. Remington, Pure; Friar Kimball, Sinclair; Maurice "Dutch" Meisse, Phillips; Joe Wheller, Stanolind; Wil Knight, Union; Curtiss McHorse, Shell; and Joe Hemphill, Magnolia

Left: "Big Boy" Love (l) and L.E. "Preacher" Ridgway at Flora in the 1940s; *above:* Dave and Olita Harrell (1927), both prominent geologists arrived in Jackson in the late 1930s, to become permanent residents.

(These three photos are Heidelberg Field, Colliers Magazine, 1945.) *Above, left:* Village storekeeper, B. C. Burns, became wealthy with oil royalties, in conjunction with a trucking and oilfield supply business; *above:* The luckiest, happiest man in Heidelberg was Harry Eddy, who, down and out in 1940, became one of its richest men; *left:* Many very poor citizens, like Mrs. Effie Husband, were among those made rich by the oil boom. Three good wells were drilled on her farm.

Eastman, Gardiner & Company office building at Laurel, 1932. The company would eventually become the Central Oil Company.

Six of the oil scouts of 1945; Francis, Stevens, and Michael became permanent residents of Jackson.

Carthage Point Field mid-1940s. (l to r) Bud Norman, Bill Vaughey, Blackey Blackburn, and Emmett Vaughey

Union Producing Company was not satisfied and elected to drill a second well approximately 12 miles southeast of the McCorvey dry hole. The location was near the east bank of the Tombigbee River at a higher structural position on the Hatchetigbee Anticline.

A contract was made with Harry I. Morgan Drilling Company for a "turnkey" well to 10,000 feet. The well began drilling May 31, 1941 and reached 10,000 feet in August: "Harry I. Morgan completed his 10,000 foot contract (probably with sigh of relief) on Union Producing Company's No. 1 Waite" (Dixie August 21, 1941). Union Producing continued drilling and eventually reached 12,399 feet on October 20, 1941. This well shattered depth records for Mississippi, Alabama, and Florida as no previous well had been drilled much below 10,000 feet.

This was the first well in the southeastern states to penetrate Jurassic sediments. Deeper formations encountered were Lower Cretaceous, Cotton Valley, Haynesville, Smackover, and Norphlet. The well bottomed in Louann salt. The scientific information gathered from this well was invaluable for future deep drilling in the southeastern states.

Bob Hearin recalls that, after the well was plugged, John Ivey, chief geologist for Union Producing Company, called a meeting of geologists and landmen in Shreveport. The accepted theory at the time was that oil migrated very slowly and some structures had not had time to "fill up." Ivy said, "Well, we're ten million years too soon." And a fellow in the back of the room, a landman, spoke up, "Well, hell, let's pay rentals till it gets there."

Humble

Another significant development was a decision by Humble Oil & Refining Company to concentrate in Alabama, possibly because of a deal between Humble and its Standard of New Jersey twin, Carter Oil Company. Humble was to operate only in counties falling generally south of Highway 80 in Mississippi and Carter was to stay to the north.

Humble took some two million acres of leases in southwest Alabama and a portion of the Florida Panhandle. Included in these leases were all of the Mobile County school lands (16th Sections), leased from the school board for a total bonus of $48,922. Eventually, leases were taken throughout Mobile, Washington, Baldwin, and Escambia counties.

Humble moved in a number of geophysical crews and rented all available houses and apartments in Evergreen, Alabama, to accommodate these crews.

The company then moved its district offices from Hattiesburg to Mobile for a period of time.

Other Alabama Activities, 1939–41

Other parties active in southern Alabama, primarily Baldwin County, were Sun, Gulf, Sinclair-Prairie, Ohio, and independent J. Brian Eby.

One of the first deals made by Vaughey and Vaughey, after becoming independents, involved a 30,000-acre block they had assembled in Clarke and Monroe counties, Alabama. A Dallas independent took the deal, agreeing to drill a 5,500-foot test, which later proved to be dry. The Vaugheys also took a block at Gilbertown in Choctaw County for Carter Oil Company.

Other activity in Alabama during this period involved a few shallow wells in the Black Warrior Basin. Geologist H. D. Easton promoted several wells in northern Alabama, some of which were financed by Seaboard Oil Company. Phillips Petroleum Company also conducted seismic work in Mobile Bay in early 1939 before moving its crews to the Mississippi Sound.

By the end of 1941, the National Oil Scouts Yearbook reports, a grand total of 275 wells had been drilled in Alabama since its first test in 1893.

CHAPTER 30

Florida, 1939–1941

Activity in Florida picked up in the years 1939 to 1941. A depth record was established in 1939, when Peninsular Oil and Refining Company drilled its No. 1 Cory well to a depth of 10,006 feet in Monroe County. Robert B. Campbell, president of the company, indicated that $200,000 had been invested in the venture, no doubt Humble's money.

Gulf and Benedum-Trees completed their geophysical investigation of the Collier Company's lands and decided to lease. However, the terms of the lease offered by Collier Company were too stringent for Gulf. Benedum-Trees, not willing to give up, worked with Peninsular to negotiate a deal with Carnes Collier of Miami. Peninsular, being a subsidiary of Humble, passed the deal on to its parent company. Humble was assigned 1,053,600 acres of the Collier Company lands in Charlotte, Lee, Hundry, and Collier counties and proceeded to survey the area with magnetometer, gravity, reflection seismograph, and coredrill.

Humble absorbed Peninsular during 1941 and subsequently stationed scout Joe Heick in Tampa. North of the Humble play, Sun Oil Company had been conducting seismograph in Collier, Henry, and Lee counties near Fort Walton, for several years. At the beginning of 1939 the company sent its top lease buyer, J. C. Vaughn from Jackson to the area. Vaughn soon had 2.5 million acres under lease for Sun, later reduced to 1.5 million. Soon, Don Monroe was stationed at Tallahassee as district geologist.

In late 1939, William G. Blanchard prepared to drill his Everglades lease at a site along the Tamiami Trail, west of Miami. He and J. L. McCord purchased a drilling rig and moved it to the location. The well was spudded January 4, 1940. A shallow gas zone was hit at 1,300 feet with sufficient gas coming up with the mud to burn a flare.

After the flare had burned over two and a half months, Blanchard sought money to drill a second well. To demonstrate to potential investors the merits of a second test, Blanchard built a pipeline from the well to the Tamiami Trail, where he installed a domestic gas range and other gas-fired apparatus in a trailer.

Governor Fred P. Cone and Governor-Elect Spessard L. Holland inspected the demonstration. Apparently the gas was insufficient to develop a commercial venture, and the second well was not drilled.

Also in 1940, Dr. Edward A. Hill set pipe on his Cedar Key No. 2 well after finding slight oil and gas shows in the Eutaw. Nothing developed from this.

On March 19, 1941, Hermand Gunter, state geologist for Florida, gave a paper to the Mississippi Geological Society at the Edwards Hotel in Jackson on Florida geology. A Dixie Geological Report of July 3, 1941, stated:

> The possibility of having some activity in Florida before this year is over is very good. Between the Gulf Refining Company, Humble Oil & Refining, Magnolia Petroleum, and the Sun Oil Company there is approximately 2,500,000 acres under lease now. . . .
>
> The Humble is shooting in the Everglades, but the results are worthless. However, they are obligated to spend $50,000 in exploration work, so it looks like core drilling or slim-holing is the only solution. (Dixie, July 3, 1941)

Humble, Sun, Gulf, and Sinclair-Prairie were checkerboard leasing on the Ocala Uplift. By the end of 1941, 80 wells had been drilled in Florida with no success. The deepest was 10,006 feet.

CHAPTER 31

Preparations for War

On September 1, 1939, Nazi Germany stunned the world by invading Poland. The Germans' new "blitzkrieg" warfare tactics used a large force of tanks and other motorized vehicles to overrun opposing forces. Petroleum was to play a major role in the pending war.

Some 85 percent of military ships burned oil, and an army division required 50 times more horsepower than it did in World War I. The United States, although neutral at the time, began to canvas its petroleum supplies and organize for possible future conflict. In 1939 the United States had 60 percent of the world's crude oil production; the Soviet Union had another 17 percent. These two countries would supply the Allies.

As the war in Europe expanded, Germany captured the Galician and Romanian oil fields, along with a large petroleum reserve stockpiled by the French. However, Germany relied on production of synthetic fuel, mostly from coal, and severely curtailed the use of petroleum by the civilian population.

The United States launched a national defense and rearmament program in early 1940, and passed the Lend-Lease Bill in March 1941. On May 27, 1941, President Franklin D. Roosevelt declared a state of "unlimited emergency."

Army Camps

Construction and expansion of military bases began throughout the country. In the southeastern states the cut-over timberland bordering the Gulf of

Mexico offered a particularly attractive habitat for new army camps. Having been denuded of its original stand of virgin pine some twenty years earlier, this land was covered with scrub brush, scraggly pines, and new growth. It appeared to have very little value to most people, but it was ideal for military training and maneuvers.

Harold Ickes

Recognizing the importance of petroleum, President Roosevelt appointed Harold L. Ickes petroleum coordinator for national defense on May 28, 1941. This did not sit well with many members of the oil industry, as their past contacts with Ickes had not been favorable.

Ickes writes, about his first meeting with industry representatives in June 1941:

> The atmosphere that prevailed as we forgathered on that summer day back in '41 was pretty much the same that one might find just before a mutineer was to be strung up to the mizzenmast. I was held to be a very bad character indeed; an unhealthy influence on our national life; one who went about looking for industries that he could clamp chains on. . . .
>
> Wasn't I one of the more toxic of the New Dealers? Didn't I look with suspicion on anyone who made a profit? Didn't I believe that Government should rule business with a blacksnake? Wasn't I the "so-and-so" who had tried to take over the oil industry back in 1934? And, finally, hadn't I aspired to be an "Oil Czar"? (Ickes, 1943)

To offset their objections, Ralph K. Davies, vice president of Standard Oil of California, was made deputy coordinator with equal power. On June 19, 1941, the Office of Petroleum Coordinator (OPC) held a meeting with 1,500 oil men and proposed a partnership between government and industry.

In July 1941 Ickes appointed a staff of specialists along with several other oil men as directors of research, production, refining, transportation, and marketing. Everette Lee DeGolyer was made director of conservation. A central office with five district offices and various suboffices were set up to cover the entire United States. These positions were filled by people with practical experience in the petroleum industry. Some 75 percent of the salaried staff members were recruited from oil companies, many at considerable reductions in pay.

The Petroleum Council for National Defense was established on September 28, 1941, with top managers from major oil companies creating a 26-

member executive committee. All served without pay. Their first meeting was scheduled for Monday, December 8, 1941.

On December 7, 1941, the Japanese attacked Pearl Harbor, and the United States was at war. The war effort placed strict controls on the civilian population and business:

> But the mood of the public in the wartime crisis was that all you had to do to increase oil supply was merely to twist a valve, and consequently the agency's program of augmented exploration and the increased drilling of "wildcat" wells, to discover new fields, was hampered both by military and public misconception. And when the public rebelled against gasoline rationing, considering the shortage of oil "phony," the oil industry itself—not the government—paid for the advertising campaign to persuade the public to recognize reality. (Tinkle, 1970)

The decision to use experienced oil personnel to administer the supply of petroleum in the event of war proved to be extremely wise. Instead of curtailing drilling and exploration, and furnishing increased crude demand by increased allowables, the Office of Petroleum Management (OPM) mandated that no oil wells should be drilled on spacing less than 40 acres. This would reduce steel requirements to 60 or 70 percent of that normally used.

An inventory was made of all tubular goods, valves, fittings, and other oil field equipment. Materials were to be allocated only to "meritorious projects." Wildcatting was encouraged and given priority numbers higher than development wells. Only necessary wells were drilled and no duplication pipelines were given priority.

State regulations were subject to federal approval. Each state was given a production quota with Mississippi's set at 60,400 barrels of oil per day (bopd). At the time, Mississippi was already producing 75,000 bopd.

The 1939–41 period ended with great uncertainty about the changes to be brought about by the war. Mississippi was listed among the oil-producing states, having gone from no ranking to tenth place during these three years. It was expected that Alabama and Florida would soon join the top ranks.

PART IV

The Boom of the War Years, 1942–1945

CHAPTER 32

The United States Enters World War II

Through the 1930s the Japanese had become progressively more militant. After they invaded Manchuria in 1931, the war with China was escalated to full scale by 1937. The powerful Japanese military machine had one fatal weakness, it was almost completely dependent on imported petroleum, about 80 percent of it from the United States.

The Japanese leaders, being very aware of their reliance on imports from the United States, and, the growing U.S. dissatisfaction with their military aggressiveness in China, were determined to end this vulnerability. Nevertheless, the United States continued to supply oil and other material to Japan into the summer of 1940, even though Hitler had overrun much of Europe during the preceding year.

It was not until the National Defense Act was passed in July 1940 that imports to Japan were limited. On September 26, 1940, the United States stopped exporting iron and steel scrap to Japan, but not oil. The Japanese began to create huge stockpiles of gasoline and other petroleum products, sensing what was to come.

The obvious solution to Japan's problem was to take the Dutch East Indies. The huge oil fields there would assure Japan of a plentiful supply of oil. They were fearful, however, that such a bold move could bring reprisal attacks from the American fleet at Pearl Harbor. To insure against this, Japanese planes carried out their surprise attack on December 7, 1941, to neutralize the U.S. naval power. The Japanese planes, fueled by gasoline from American oil wells and refineries, devastated the Pacific-Fleet in the Pacific (Yergin, 1991).

The nation was stunned by the Pearl Harbor atrocity. With often misleading news being furnished by radio and newspapers, the average American knew little about Japan or events leading to this aggression. As the Christmas of 1941 was at hand, American stores were filled with toys made of crimped tin, papiermach'e, wood, and bamboo, each stamped "Made in Japan." To most Americans, "Made in Japan" was synonymous with goods of inferior quality that sold for pennies. "Made in U.S.A." meant top-quality, well-built, but more expensive products. Though I was only 12 years old on December 7, 1941, the attack on Pearl Harbor is still a vivid memory. In Palestine, Texas, people received the news with disbelief. No one was sure what was to come. There was a suppressed undercurrent of excitement. Rumors were circulating that the "Japs" were on their way to invade California. Radios were left on all day to catch the latest development. As days passed, daily routines were gradually resumed.

As 1942 began, a sense of foreboding swept the country. The United States declared war on all the Axis countries: Japan, Germany, and Italy. Initially the war put a damper on new drilling projects, even though a tremendous need for petroleum was anticipated. Pipelines and refineries would be strained to the limit. Crude oil supplies were adequate, but no one knew how long the war would last or how much demand would be placed on petroleum. Soon a 35-mile-per-hour speed limit was mandated by the federal government. Gas rationing was started, with coupons issued to each driver. Wage and price controls governed the civilian sector.

The federal government continued to encourage exploration for oil. It was anticipated that new reserves would be required in the future to supply military and domestic needs. Industry came under government controls for the duration of the war. The government became dictatorial, even questioning the right of each state to manage its own oil business. Of primary concern was unnecessary drilling in view of the short supplies of steel and other material. The government decreed that each oil-producing state should legislate a set of conservation statutes along guidelines established by the Office of Petroleum Coordinator. Everette L. DeGolyer was the Petroleum Coordinator official for the Gulf Coast including the southeastern states. The legislatures of oil-producing states rushed to comply.

CHAPTER 33

Mississippi, 1942–1945

The Legislature

At the beginning of 1942 Mississippi oil operators were concerned with bills pending in the state legislature. These included a severance tax to be levied against oil production and the federal decree for conservation legislation.

The legislators worked with oil representatives to establish the conservation rules. At first there appeared to be accord on the issues:

> The Controversy over spacing, production, and conservation in Mississippi is being straightened out satisfactory to the petroleum coordinator. The Attorney General, with the assistance of others, is drafting a conservation bill which is based on Arkansas Law. . . . The OPC will not consider any exception in Mississippi. (Dixie, January 15, 1942)

> Events have occurred in other states that might throw a little light on what can be expected. The Illinois Oil & Gas Conservation Division was ordered by the Federal Government to suspend issuance of all drilling permits [10-acre spacing]. In Arkansas, during the absence of the Chairman of the State Oil & Gas Board, several 10-acre drilling permits were issued, but they were immediately revoked upon his return. (Dixie, January 8, 1942)

> Dr. E. DeGolyer, Director of Conservation of the Office of the Petroleum Coordinator, addressed a joint session of the Mississippi Legislature Tuesday afternoon, January 20th. Dr. DeGolyer urged the state to "take up the job of managing production of oil and gas so that it will be properly conserved and the physical waste avoided." He stated that in his opinion the

Tinsley Field's total recovery had been reduced by 40 percent due to spacing and unlimited withdrawals that have occurred and that this situation was due to the fact that Mississippi does not have a Conservation Law. He also stated that due to the present National Emergency that if Mississippi did not properly manage its oil production that the Office of Petroleum Coordinator [OPC] would use all resources in its power to see the industry was properly regulated. (Dixie, January 22, 1942)

Perhaps DeGolyer's threat did not sit well with the "Old South" feelings of the legislature:

The proposed Conservation Bill was introduced into the senate on Thursday afternoon, February 26th. By vote of 24 to 21 it was "tabled," which is the end of the measure so far as this session is concerned . . . The senator who led the opposition on the floor referred to the OPC men who have recently been in Mississippi as "Office Boys." The outcome of this bill has been nothing less than nose-thumbing at the OPC. The strong opposition to this bill cannot be accounted for by other than politics. (Dixie, March 5, 1942)

The general political set up in Mississippi seems to be more of ox-cart days than even horse and buggy days. (Mississippi has a Sunday Blue Law, and on Tuesday afternoon, March 11th, after several "Billy Sunday's" tirades, the House of Representatives voted down Sunday movies even though the galleries were packed with soldiers from the Jackson Air Base.) (Dixie, March 12, 1942)

Thus, Mississippi failed to pass a conservation bill in 1942, despite pressure from the federal government.

While the fight on conservation was going on, the Governor was more concerned with taxing the oil business. Governor Paul B. Johnson addressed the Mississippi Legislature in January 1942:

Now we have discovered another resource just as valuable as timber. The exploiters are here. They have purchased, for a nominal sum, the land on which this oil is situated. If you do not pass a severance tax and enforce it, the time is not far distant that we will curse the day we permitted this great natural resource to be taken from us as we permitted the timber to be taken. You can rest assured that the men who come in here to develop the oil are not, as a rule, interested in anything except all the money they can get out of the oil. When they have depleted the fields of the oil, they will return whence they came and we will regret that we did not collect a reasonable tax. (Dixie, January 15, 1942)

The governor learned that his proposed seven percent tax on oil was not going well with the legislature. In desperation he made another address:

> We have had a notorious lobby for the oil interest. So rotten has it been that you can smell the oil around where they have been. There is a great deal of loose talk about this situation. Some of it is not very complimentary. Some of the oil lobbyists have made statements that we could hardly believe to be true about the situation at the capitol. (Dixie, February 26, 1942)

The governor's bill was killed before the legislature adjourned.

The 1942 legislature did enact and the governor approved a bill to clarify the law in regard to oil and gas leases on school land in Mississippi. This was HB668 which provided that

> the Board of Supervisors, upon approval of the County Superintendent of Education, of each county are hereby authorized and empowered, in their discretion, to lease 16th Section lands in their respective counties. . . . However, that said school lands shall not be leased for oil, gas, and mineral exploration and development for less than $1 per acre per annum, . . . for a primary term of more than six years. (Dixie, March 26, 1942)

The bill went on to provide that a royalty of one-eighth be paid for both oil and gas, one-tenth on other minerals and 50 cents per long ton on sulfur.

The Mississippi State Legislature met only every other year. At its next biannual session, the legislature followed Governor Bailey's request and, on February 29, 1944, the governor signed HB23, which placed a tax of six cents or six percent on crude oil, whichever was greater.

The legislature again failed to pass a conservation bill during the 1944 session, after much bitter debate.

Materials and Labor

In 1942, the government was balancing domestic industry with the demands of the military. Several selective service boards ruled that all workers on more than one hundred oil related jobs who were married prior to December 7, 1941, were to be classified as 3-B. This list of jobs covered almost the entire petroleum and natural gas industry from exploration to refining. The 3-B classification that almost assured that one would be allowed to serve in a civilian job rather than in the military. (Dixie, August 27, 1942)

This did not mean that everyone was exempt, however:

> Local Draft Boards are hot on the heels of a lot of the local oil men. L. R. McFarland, Magnolia's Double-District Geologist, and recently elected Vice-President of the Mississippi Geological Society, has received his induction papers and is to report on October 21, 1942. (Dixie, October 15, 1942)

McFarland was but one of the many oil people being called into the service. The ranks were rapidly being thinned in the oil business. Those remaining behind were doing double duty.

> The labor situation for drilling contractors in Mississippi is becoming very serious. Harry Morgan, contractor on Love Petroleum Company's test at Flora in Madison County is forced to work two twelve hour shifts. Sam Cook, Contractor for Carter Oil Company's Bridgeforth well at Pickens intends to stack his rig as soon as he finishes this well because of the labor shortage. Heretofore, the biggest headache has been materials, and the labor problem is a recent addition which is due not only to Selective Service [the draft], but to curtail the work, which forced the workers to turn to steady defense work. (Dixie, October 29, 1942)
>
> The old law of "supply and demand" is quite far reaching in its effect sometimes. It will be interesting to wait and see if the unwritten Gentlemen (?) agreement among various Major companies not to hire a Geologist employed by another Company will stand all the way through the manpower shortage.
>
> Adjustments of continued depression wage scales have already occurred in most cases, but there is not a black market on geologists yet. (Dixie, March 18, 1943)

Among the many who served in the military were Wil Knight, J. A. Morgan, and Bob Hearin. Emmett Vaughey attempted to join the navy, but was turned down due to his age.

The industry also felt the pinch from shortage of materials:

> Trucking of materials is a very serious bottleneck. Repair parts for trucks in some cases are harder to get than new tires.
>
> Some of the trucking contractors almost refuse to send their trucks to swampy locations. Recently it is taking a minimum of two weeks to move materials and rig up on several of the wildcats that are now being drilled in Mississippi. (Dixie, September 24, 1942)

CHAPTER 34

Drilling Resumes

Almost all of the wells at Tinsley had been drilled by the end of 1941, before the war began. In January 1942 production at Tinsley had reached 90,000 barrels per day. The federal government's production quota for states, limited Mississippi to 55,600 barrels. However, many other states, such as Illinois, fell far below their quotas so no action was taken against Mississippi for the over production.

Despite the lack of drilling at Tinsley, a number of interesting deals were made.

The Sohio Corporation acquired interest in Tinsley, purchasing the W. G. Ray Walker lease with three wells for $115,000 plus a one-eighth override. Sohio then purchased E. C. Johnson's barge line operating north on the Mississippi River along with his producing leases at Tinsley and his refinery at Vicksburg. The total consideration involved in this deal was reported to be over $750,000, the largest oil deal ever consummated in Mississippi's oil industry history to that time (Dixie, December 17, 1942).

Sohio then opened a district pipeline office in Jackson, Mississippi, and purchased the Southern Pipeline Company from Sylvan Producing Corporation of Mt. Vernon, Illinois. This purchase gave Sohio some of the gathering system in Tinsley. The company also bought Sinclair's loading racks at Tinsley and Pickens (Dixie, June 3, 1943). Sohio was preparing to build a pipeline after the war and thereby control most of the Tinsley crude.

After reaching peak production in early 1942, the Tinsley Field began a slow decline, which would continue for the next 50 years. The operators began to concentrate on increasing the price being received for the crude. H. L. Hunt and the Standard Oil Company of Louisiana applied to the Office of Price Administration for a price increase. At the start of the war, the Tinsley crude price had been frozen at $1.00 for 40-degree gravity with a two cents price differential per degree downward. This gave an average price of about 90 cents. The base price was raised to $1.18 on May 22, 1943.

Pickens

Only five wells had been completed at Pickens at the beginning of 1942, even though the field was over two years old. The completions of Buzz Morgan's well in the Sharpsburg area caused drilling to pick up. Both Phillips and Carter began new wells. All of the early wells were plagued with the same problem of producing water with the oil. However, in 1943 an event of geological significance took place. A fault was evident on the electric log of the Phillips Petroleum Company No. 1 Whitworth well. This fault would eventually be proven to lie along the northeast side of the structure and would be the "trapping" feature of the field. Structurally higher wells could be completed water-free."

By the end of 1942, nine wells had been completed in the field by four operators: Sinclair, Phillips, Carter, and Buzz Morgan. Daily average production had risen to 1,600 barrels per day. Other independents entered the picture. Tip Ray, Lloyd Spivey, and Theo Denkins, all of Canton, along with Bill Vaughey of Jackson, consummated a farmout deal with Carter Oil. John D. Golson of Jackson agreed to drill the well for a one-quarter interest plus $10,000. The No. 1 Wilson made an excellent oil well and extended production in the field half a mile to the northwest.

Effective September 1, 1943, Carter Oil Company became the purchaser of Pickens Crude Oil. Up until that time oil had been shipped by rail to St. Elmo, Illinois. Carter opened a district production office in Canton, with R. W. Hurt as superintendent.

Vaughey & Vaughey next drilled its No. 1 Dixon well, the highest structurally in the field to that date, which would ultimately produce a million barrels of oil. The well was drilled on a farmout from Sinclair, granted over the objections of Don Reese, its district geologist who wanted Sinclair to drill the well. Reese said, "What's wrong with them [the management]! We ought to drill this thing. His boss, Jack Gordon, said, 'Don, let sleeping dogs lie' " (Vaughey & Reese, March 11, 1987). While the well was being drilled, Bill Vaughey called Reese and asked him to help them locate the Wilburn Sand. Reluctantly, Reese said,

"O.k., I'll come up there". . . . They cut the fault in the core, it was perfect. I said, "you guys take another core and you will have oil sand in the morning." So we went to Tip Ray's lodge [and made merry]. We got up and went out there the next morning and there it was. . . . Isn't that something. We hit it right on the button. (Vaughey and Reese, March 11, 1987)

We came out of this discovery [Pickens] with our first Mississippi production. We got two wells, one a farmout from Sinclair and one a farmout from Carter Oil. One of the wells, No. 1 Louise Dixon, turned out to produce a million barrels of oil. (E. A. Vaughey, March 11, 1987)

Don Reese and the Vaugheys became close friends and entered into many successful business deals after the war.

The Pickens Field lies along the Illinois Central Railroad:

We [Vaughey & Vaughey] had to notify them [ICRR] before we could try to complete a well. They couldn't say no, but we had to let them know. So, Roy Whiteman, who worked for us, called Illinois Central and told them that we were getting ready to complete a well there. It was right by their track, and so forth. She [a secretary] said, "We don't have a track anywhere near that." He said, "Lady, if I was you, I would get on [the phone] to the Vice President right away, because one of your trains came by here like a bat out of hell."

By the end of 1945, a total of 41 producing oil wells had been drilled at Pickens. The field that started out with a doubtful future was now producing over two million barrels per year, with a cumulative production of almost seven million barrels by the end of 1945. Sinclair missed its chance to become a significant producer in Mississippi in these early years by not recognizing the potential of Pickens Field.

Vaughey & Vaughey went on to drill several more locations in Pickens including the No. 1 Sally Ray, the No. 2 Edna W. Garrett, and the No. 1 Covington and Edgar. Their success gave the Vaughey & Vaughey partnership substantial income which enabled it to become one of the most successful independent oil operators of Mississippi.

CHAPTER 35

Drilling Trends in Mississippi

In 1940, a shallow Wilcox field was discovered in northern Louisiana approximately 50 miles west of Natchez, Mississippi and north of Alexandria, Louisiana. This was the Olla Field, which proved to be a very large, prolific, shallow oil field. Depth of the production was only around 2,800 feet. Thereafter, the Wilcox was of particular interest in southeast Mississippi and for a time it was referred to as the "Olla Trend."

The first mention of the Wilcox Trend by the Steffey Report was on January 28, 1939:

> It was disclosed at a meeting of the Houston Geological Society by Geologist John Todd, an authority on the formation as follows: "The World's newest, largest, and richest oil formation, a 1,000 mile long and 100 mile wide Sparta-Wilcox trend, had been found in Texas, Louisiana, and Mississippi. Its true value in dollars is incalculable. . . .
>
> This formation, Sparta-Wilcox, is found at a depth of approximately 9,000' on the southern edge of the strata and at around 5,000' on the northern edge. It was stated that the formation at greater depths may be even more prolific but commercial wells below the 10,000' level are too costly to warrant research under present drilling conditions. . . .
>
> So far, the Sparta-Wilcox trend is producing oil at Eola [Olla], La; Ville Platte, La; Cheneyville, La; Spurger, Texas and Cleveland, Texas.
>
> Due to discovery of above mentioned "Sparta-Wilcox" trend, oil activities reached an all-time high, in Wilkinson County, Mississippi, with exploration headquarters being established by several companies in the

town of Woodville. It is reported that leasing and royalty buying has increased sharply the past two weeks. . . . NOTE: It looks to us as though they were "Fixin to Flush a Covey."

The Dixie Geological Service reported two years later:

> The Wilcox Trend play in Louisiana has now extended eastward to the Mississippi River, but so far volume buying has not started in the western Mississippi counties. However, several buyers of quietly accumulating acreage in Claiborne, Jefferson, Adams, Franklin, and Wilkinson Counties, Mississippi. . . .
>
> Practically all of the wells that have been drilled in this area did not test all of the horizons that are now producing in Louisiana. In light of what has been learned in Louisiana, it seems that if any well which is to be drilled in the Wilcox Trend should be carried in through the entire Wilcox section. Before Olla, the top of the Wilcox was the only part that was watched. . . . (Dixie, September 11, 1941)

> In regard to Wilcox tests, it has come to the point where it is hard to say just what penetration can be considered to have tested the Wilcox. The Atlantic Refining Company has recently obtained production in Beauregard Parish, Louisiana, 2500 feet in the Wilcox. (Dixie, September 25, 1941)

This interest was intensified in 1942 when the California Company found Wilcox production immediately across the river from Mississippi in the Lake St. Johns Field, Concordia Parish, Louisiana. This was at a depth of approximately 4,600 feet. In February the Fisher No. 1 Lambdin well in Adams County reported a show of oil at 5,690 feet in the Wilcox. Even though the show was not of commercial value, this was the first well in Mississippi to encounter a "Louisiana Wilcox show."

In April 1942 the Magnolia Petroleum Company drilled its No. 1 Pidgeon well in Warren County, Mississippi, near Vicksburg, which blew out at 1200 feet from a Sparta gas sand. The Sparta lies immediately above the Wilcox.

Black Warrior Basin

Some companies were still interested in northern Mississippi:

> Ohio Oil Company covered the Black Warrior Basin with the new gravity before 1940. They dug their largest. . . . anomaly, Ohio #1 Cayle in southeastern Chickasaw County, and checking with seismograph, finding no structure-folded their tent. (Phillips, 1989)

Carter Oil Company took Ohio's place, becoming quite active in northern Mississippi by 1939. The company did considerable seismic work in the Amory and West Point areas of Monroe County. Bob Phillips reports:

> Carter Oil did a careful and good seismic survey of the Basin. First, checking Bill Jenny's Ridge and found it to be a fault, worked along both sides of the fault. They mapped Muldon, Aberdeen, Mayben, Siloam, and Corinne. They dug Muldon and with no mudlogger, plugged it in 1941. (Dixie, June 19, 1941)

Phillips was referring to the Carter Oil Company No. 1 J. T. Sanders well in Monroe County, drilled in the summer of 1941. Even though this well was abandoned as a dry hole, it would later prove to be the discovery of the Muldon Field.

Carter also drilled a test well in Chickasaw County, Mississippi, on the T. G. Abernathy farm. An attempt to complete the well failed. Nevertheless, the sand became known as the "Abernathy Sand." Dixie reported,

> The Carter [Oil Co.] seems to be somewhat disgusted. Instead of moving their strat rig a few thousand feet or a few hundred miles, they are moving several thousand miles. The rig is being moved to South America. (Dixie, April, 14, 1942)

However, the Dixie Report of February 18, 1943, stated, "The Carter Oil Company No. 1 J. T. Sanders well in Monroe County, is reported to be making some gas through the plug." This very important note in the scout report which was apparently overlooked as unimportant by the oil community.

CHAPTER 36

The Fabulous New Discoveries

With the uncertainties created by the war effort, drilling was greatly curtailed in the southeastern states during 1941 and the first part of 1942. No new discoveries were reported in 1942, but by the end of that year, drilling was picking up. Oil companies drilled many structures that had been mapped and leased during the late 1930s and early 1940s. The federal government encouraged this exploration.

Twelve new discoveries were made in Mississippi during the last three years of the war, 1943 to 1945, in spite of the limited material and workers available.

The California Company Finds Brookhaven

Immediately across the river from Mississippi in Louisiana, the California Company had found some very prolific Lower Tuscaloosa production in addition to Wilcox production at Lake St. John. The depth was 9,500 feet. The company had completed two wells in the field by early 1942.

In the fall of 1942 The California Company began to drill a test on a block at Brookhaven, Mississippi. This structure had been found on the company's original seismic line shot in 1937. The No. 1 G. T. Smith well reached a total depth of 12,229 feet with oil and gas shows logged in the Lower Tuscaloosa. Pipe was set at approximately 10,600 feet to test these shows. To everyone's surprise, the well produced oil rather than gas. The completion was official on May 31, 1943, with production from perforations between 10,138 and 10,322 feet in the Massive Sand of the Lower Tuscaloosa formation. It was necessary for the California Company to install a hydraulic pump on the well to generate a reasonable rate of production. The well pumped 153 barrels per day of 25 degree gravity oil.

This was an unusually deep completion for the United States at this period. Production at 10,000 feet was usually gas rather than oil. The depth and the low-gravity oil tended to make the company regard the discovery as insignificant:

> The area is definitely cool and it is going to take more than the well has shown so far to create much royalty trading. Close in minerals can be bought from $25 to $50. (Dixie, January 7, 1943)

Sohio began purchasing the Brookhaven crude at 88 cents per barrel. Prior to this, Danciger had purchased the crude to use as fuel on its rig drilling a well for Helis in Lawrence County, Mississippi.

The California Company's second try, No. 1 Hattie Smith, only half a mile to the east, was a dry hole. Upon missing the pay sands, it drilled to a depth of 12,747 feet which almost tied the depth record for Mississippi. This failure led the California Company to further discount the value of the Brookhaven discovery. The third try, the No. 1 Grandberry Smith a mile to the north, was also dry. The California Company gave up. There was no more drilling at Brookhaven until 1945.

However, in the oil business there is always an optimist willing to try another well. Roeser and Pendleton, Inc., of Fort Worth, Texas, obtained farmouts from the California Company and the Sun Oil Company on several leases. The farmouts required a test well to be drilled approximately one and three-quarters mile south and slightly west of the discovery well. The Roeser and Pendleton No. 1 W. L. Case well surprised everyone when it flowed 240 barrels of 37-degree gravity oil per day. Completion was official in September 1945.

As it turned out, Brookhaven was a "bald top" structure with some Lower Tuscaloosa oil sands absent over the crest. Most of the development drilling in the field would take place after the war. The field would eventually have more than 150 wells that would produce some 75 million barrels of oil and 350 billion cubic feet of gas, not too bad for such a dismal beginning.

Several independents would benefit from Brookhaven along with Roeser and Pendleton. One was Vaughey & Vaughey:

> In 1943 the California Company discovered the Brookhaven Field and developed it. At the time I was taking the block for them [years before],

there was a piece of land owned by the Federal Land Bank, which they did not want and which we bought. It turned out that it also produced and that was our next oil in Mississippi. (E. A. Vaughey, March 11, 1987)

Gulf's First Success—Eucutta

At the beginning of 1943 all of the oil fields that had been discovered in Mississippi were in the western part of the state. Gulf had spent large amounts of money in eastern Mississippi, in areas near Waynesboro, Laurel, Hattiesburg, and Lucedale. Many large-gravity anomalies had been located that were suspected to be salt structures. Gulf decided to make a move:

> Since the first of this year, there has been talk about the Gulf Refining Company having a budget for an extensive Wildcat drilling program in Mississippi this year. In support of this rumor is the fact that Gulf is buying outstanding leasing interest and leasing additional edge acreage on all of their blocks in Jones, Wayne, and Walthall Counties. (Dixie, March 25, 1943)

Gulf made three locations, one in Greene County, one in Jasper County (the No. 1 Helen Morrison), and one in Wayne County (the No. 1 Aden-Davis):

> It has been reported previously from several sources that Gulf has appropriations for twelve wildcats in Mississippi in 1943. Now it is reported that this has been raised to seventeen . . . but half of these will be on shallow tests on top of salt domes . . . which could be classified as sulfur tests. (Dixie, April 15, 1943)

> It remains yet to be seen just how good Gulf Oil Corporation's Gulf Research Department's Geophysical Department is at finding oil in Mississippi, but they seem to be batting 1000 percent on picking Shallow Salt Domes. (Dixie, April 24, 1943)

Buzz Morgan was chiding Gulf in his Dixie Report, but the fortunes of Gulf were about to take a turn for the better. Sixteen years of intensive exploration effort, beginning with the surface-mapping crews in 1928, had been a failure for Gulf. The company had spent a tremendous amount of money in the southeastern states, both collecting geological and geophysical data and drilling many wells, all to no avail. The company had even missed out on Tinsley. Now Gulf was going to try again, probably with a great deal of misgiving.

Gulf began drilling its No. 1 Aden-Davis well 12 miles east of Laurel in

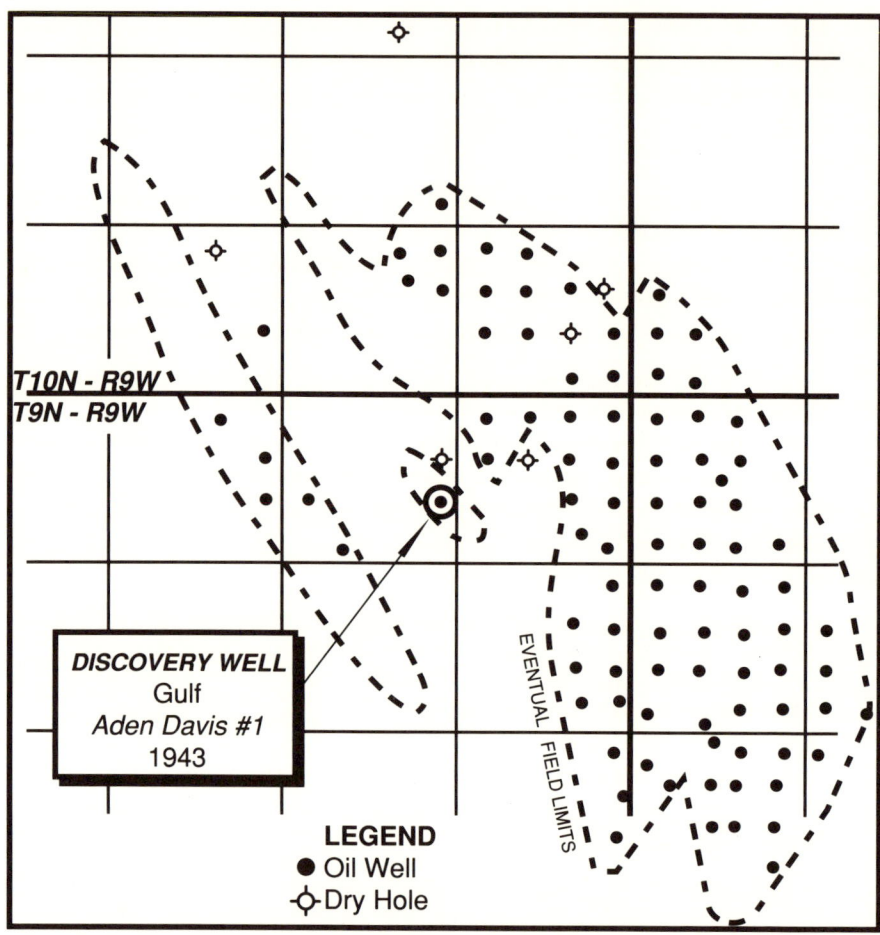

FIG. 9   Eucutta Field, Wayne County, Mississippi, 1945

the summer of 1943. After reaching a depth of 6,800 feet, Gulf set pipe to test the Lower Tuscaloosa and Eutaw Sands. Royalty in the drillsite lease began selling for $300.00 per acre. Test results were far from encouraging, however. The well produced very low-gravity crude from the Lower Tuscaloosa Sand and tended to produce water with the oil:

> Standing by the side of the tank and listening to the flow, one gets an impression not of the well flowing by heads but of plunking, kind of like something falling on a flat rock. (Dixie, August 12, 1943)

The well was completed on September 7 at 6600 feet in the Lower Tuscaloosa, pumping 63 barrels of 22-degree gravity oil per day. Although this
well was marginally commercial, it prompted Gulf to move to another location one mile to the northeast, which found a new fault segment that was
to prove to be the largest portion of the Eucutta Field. The No. 1 Stanley
found oil in the Eutaw formation. Gulf elected to take the well on to
11,055 feet, but no additional pay sands were found. The well was completed in the Eutaw at 5,200 feet. The sand thus completed became known
as the "Stanley Sand."

In the meantime, Humble had started a well on its No. 1 City Bank
Farmer's Trust Company immediately northeast of the original Gulf Davis
well. This well was completed in a new Eutaw Sand, which became the City
Bank Sand, on March 14, 1944, for 192 barrels of 27-degree gravity oil per
day. In view of the time that it took Gulf to deepen the Stanley well, Humble was given credit for discovering the Eutaw production at Eucutta.

In 1945 Gulf moved west of the original Davis well to find a new fault
trap which became known as the "West Segment." This well was completed
May 16, 1945, at 5,164 feet, flowing 240 barrels of 37-degree gravity oil per
day. By the end of 1945, 94 wells had been drilled in the field and production had reached over 2 million barrels per year. Eventually, the field would
have more than 150 wells and ultimately produce some 60 million barrels.

Love Petroleum Discovers Flora

Union Producing Company had been busy developing Tinsley since 1939
and had done very little wildcat drilling. The company began a seismic survey in Madison County, north of Flora, in late 1943, "the first exploration
work for Union in Mississippi since Tinsley-Flora-Midnight Days," (Dixie,
November 24, 1943). However, Union Producing decided to farm out the
Flora block. Love Petroleum Company agreed to drill a 9,000-foot well offsetting Union's old well which tested some oil from the gas rock in 1939.
For drilling this well, Union Producing agreed to give up one-half interest
in its block to Love Petroleum.

In 1939 the Mississippi Ordnance Plant had been built at Flora by the
U.S. government to manufacture ammunition. A thousand acres of Union's
block was put off limits for drilling in the vicinity of the plant, as a series of
buried igloo-type bunkers had been constructed to store explosives there.

In the fall of 1942 Love Petroleum drilled its first well, the No. 1 J. R.
Anderson, et al. The location was approximately 700 feet southeast of the

Union Producing Company well. Shows were encountered in the gas rock at approximately 4,400 feet. Seven-inch casing was run and perforated, but the well swabbed saltwater with a noncommercial amount of oil.

Love Petroleum, having substantial income from a lease in the Tinsley Field which was producing over 1,000 barrels per day, was encouraged to try again at Flora. The next attempt was a mile north of the old Union Producing Company well. The Love Petroleum Company No. 2 Anderson proved to be the discovery well for the Flora Field, completed on October 27, 1943. Production came from open hole at 4337 to 4367 feet and flowed 122 barrels of 25-degree gravity oil plus 43 barrels of saltwater per day. Both Love and Union Producing were discouraged by the saltwater.

In late 1943 Tip Ray and associates (the "Canton crowd") took a second farmout from Union Producing on a tract north of Love's No. 2 Anderson well. Love's well had been put on pump and settled down to a rate of 85 barrels per day with more water.

The Tip Ray well proved to be a producer similar to the discovery well. The high water cuts and low rates of production made the field commercially doubtful. One more well was drilled before the end of 1945. Other development was pretty much stymied by the Mississippi Ordnance Plant:

> It is reported that the government has men signing up mineral owners on plant property to an agreement to extend the leases, which are expiring until six months after the duration [end of war] for consideration of $1 per acre rental per year. (Dixie, December 9, 1943)

Even after the war ended there was little development at Flora for many years, and not until the 1970s would the true potential of the field be uncovered.

Cranfield

Even though the California Company was a latecomer to Mississippi compared to Gulf Oil, it was to become one of the big players in the state. Following the success at Lake St. John and Brookhaven, the company moved in a rig to drill a large structure, similar to Lake St. John, in Adams County, Mississippi.

A large anticline had been found near Cranfield by a seismic program in the late 1930s. It appeared to be a large, deep-seated salt dome some 15 miles east of the town of Natchez.

In the summer of 1943, the California Company commenced its No. 1

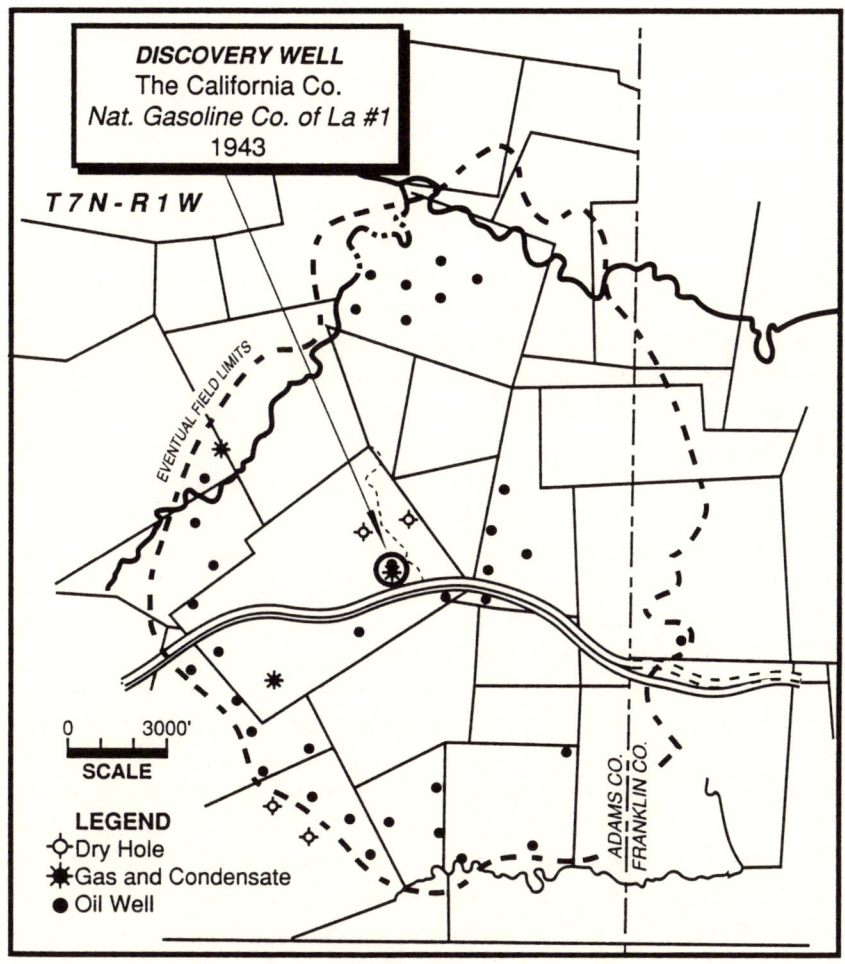

DISCOVERY WELL
The California Co.
Nat. Gasoline Co. of La #1
1943

T 7 N - R 1 W

EVENTUAL FIELD LIMITS

ADAMS CO.
FRANKLIN CO.

0   3000'
SCALE

LEGEND
Dry Hole
Gas and Condensate
Oil Well

FIG. 10   Cranfield Field, 1945, Mississippi's first gas condensate field and first
Wilcox sand production

National Gasoline Company well on the crest of the structure. Between the
depths of 4,000 and 6,000 feet, three oil sands were drilled in the Wilcox
section. The well was drilled on to 10,383 feet and seven-inch casing run.
Shows were also found in the Lower Tuscaloosa.

Brokers rushed to the area to buy leases and royalty; leases went for
$10.00 per acre and royalty as high as $60.00 per acre. Cranfield, Brookhaven, and Lake St. John made three successful wells in a row for The California Company. The O'Brien Brothers of Shreveport, Louisiana, bought a

one-quarter royalty interest under a 2,200-acre Ratcliff tract, offsetting the well to the south, for $105,000.

On October 4, 1943, the California Company's No. 1 National Gasoline Company well was officially completed producing from the "Massive Sand" of the Lower Tuscaloosa at 10,270 to 10,276 feet. The well produced at the rate of 106 barrels of 56-degree gravity distillate along with some 10 million cubic feet of gas per day. This was Mississippi's first gas condensate well. The producing sand was 85 feet in thickness, indicating a very profitable discovery.

The company also tested a Wilcox sand at 5,880 feet flowing at a rate of 280 barrels per day of 44-degree gravity oil. A second sand tested at 4,388 to 4,392 feet, and flowed 151 barrels per day of 32-degree gravity oil. In view of the lack of a gas market, a dual completion was made between the two Wilcox sands using two tubing strings. (Each sand was produced separately.) This was the *first* Wilcox well to be commercially produced in Mississippi, of the hundreds that were to follow.

The California Company drilled its second well, the No. 2 National Gasoline Company, at a lower position on the structure to find the entire Massive Sand of the Lower Tuscaloosa saturated with oil. The reservoir proved to have a "gas cap" that had been tested in the first well.

This confirmation well flowed 359 barrels of 40-degree gravity oil per day from the Lower Tuscaloosa at a depth of 10,344 to 10,391 feet. Cranfield was immediately recognized as a significant field and development proceeded very rapidly. However, Dixie reports that, "The California Company seems to be having hard luck on all of their Wilcox tests."

With the cementing techniques of the time, it was very difficult to get a good cement job on the loose Wilcox sands. As a result of poor cementing, most of the Wilcox completions were plagued with a high percentage of water being produced with the oil. The California Company decided to concentrate on the drilling of "Massive Tuscaloosa Sand" oil wells in the initial development stages of the field. This meant making locations that would fall below the gas cap. The Wilcox completions were too uncertain.

Kirby Petroleum Company purchased one-eighth mineral interest in 1,531 acres at $400.00 per acre, a very high price. In June 1944, sealed bids were taken on oil and gas leases for U.S. Forest Company Lands on the Cranfield structure in Adams County, Mississippi. H. L. Hunt was highest bidder, paying an astronomical $225.00 per acre, a total consideration of $160,000 for the lease on 714 acres. This lease was 1.5 miles south of production at Cranfield. Some 2.25 miles from the first wells, Atlantic Refining Company paid $70.00 per acre on 127 acres. These prices were staggering to the industry.

On June 8, 1944, in the Dixie Report, Buzz Morgan stated:

> As development proceeds in Cranfield Field, Adams County, Mississippi, the field appears . . . more and more to establish itself as a suitable reserve for the building of a cycling plant. The operation of a cycling plant in connection with production of a so-called "rim oil field" is one of its most practical uses. In such fields continued withdrawal of oil will result in higher and higher gas/oil ratios and the loss of considerable hydrocarbons and the wasting of large volumes of gas. Cranfield has established itself as a reservoir with an extensive gas cap and an unknown quantity of an accumulation of oil on at least its west flank.

Seeing the potentially large reserves at Cranfield, the Standard Oil of Louisiana refinery decided to lay 18 miles of 8 5/8 inch crude line to gather the oil to its barge terminal at Gibson Landing on the Mississippi River. The first oil was pumped to the landing on November 29, 1944.

In early 1945 Gulf Refining Company made a trade with H. L. Hunt. In return for one-half interest in Gulf's 1,796-acre Ella G. Lees lease on the east side of Cranfield, Hunt traded an interest in a west Texas property. Hunt agreed to drill 13 wells free for Gulf on the Lees lease, and Gulf agreed to drill two wells on the Texas 1,400-acre tract with the option to accept or reject the Texas interest. If Gulf elected not to take the Texas lease, Hunt was to pay Gulf $750,000 out of production in order to earn the interest in the Lees lease.

By the end of 1945, 33 wells had been completed in Cranfield. The field would ultimately produce some 50 million barrels of oil and 700 billion cubic feet of gas from 90 wells. This would be Mississippi's second-largest gas field.

Gulf's Second Discovery, Heidelberg

The Heidelberg structure had first been recognized by Bud Norman when he was working on the Gulf surface-mapping crew in 1929. Eastman-Gardiner had made a location on the Morrison lease in 1933, then decided not to drill. Over the ensuing years, Gulf had outlined a huge structure, using core drilling, gravity meter, and seismograph.

While continuing to drill more oil wells at Eucutta, Gulf made a location at Heidelberg on the Helen Morrison lease. In October 1943 Gulf awarded the drilling contract to W. G. Ray. The well was spudded and reached a depth of 3,830 feet by December, but the derrick collapsed. Fortunately, no one was killed, but the well was delayed while the equipment was repaired. Soon after the drilling was resumed, the Dixie Report of December 23, 1943, proclaimed:

> HEIDELBERG IS YEAR END MISSISSIPPI DISCOVERY! GULF REFINING COMPANY'S JASPER COUNTY WILDCAT OPENS NEW FIELD! WELL IS 500 FEET HIGH ON TOP OF EUTAW!

A drillstem test recovered oil and the play was on. Resident lots in the town of Heidelberg sold at $100.00 per acre. Royalty sold for $200.00 per acre within a mile of the well. The Dixie Report noted: "As is the case in most hot plays, there are some escrow artists present, but they are quickly spotted" (December 23, 1943).

A second drillstem test at 4,900 feet recovered oil at the very high rate of 116 barrels per hour on December 24. This sand became designated the "Christmas Sand" and has since proved to be the best producing zone in the Eutaw formation. The Helen Morrison well was completed on January 27, 1944, flowing at the rate of 160 barrels of 23-degree gravity oil per day. A minimum of 65 feet of oil sand was indicated in the well by electric log.

Gulf proceeded to open a production office in Laurel, Mississippi, planning to transport oil through a pipeline with a loading rack on the Southern Railroad, south of Heidelberg. Four 500-hundred barrel tanks were erected on the Helen Morrison lease.

Gulf's second well found a new fault segment that was even more prolific than the discovery well. This well had 209 feet of saturated oil sand. Reports were that "HEIDELBERG MAY PROVE TO BE MISSISSIPPI'S LARGEST RESERVE!" (Dixie, March 16, 1944). Offsetting the Morrison well, the Sun Oil Company owned the Mack Lindsey lease and commenced operations. Thanks to Jack Vaughn, Sun had a good lease position on the structure.

However, Sun was not so lucky on its No. 1 Mack Lindsey well, which encountered gas in the "chalk" and blew out at 3,800 feet. Flow was estimated as high as 20 million cubic feet of gas per day. The company was forced to abandon the well as it could not control the gas. Sun was successful on its second try, the No. 1-A Mack Lindsey.

The risk of a blowout was to plague future drilling in Heidelberg and increase the price contractors charged for drilling. The Oil and Gas Board established 40-acre oil spacing for Heidelberg, except that the town lots were allowed to be drilled on 20-acre units. Unitization of the townsite lots

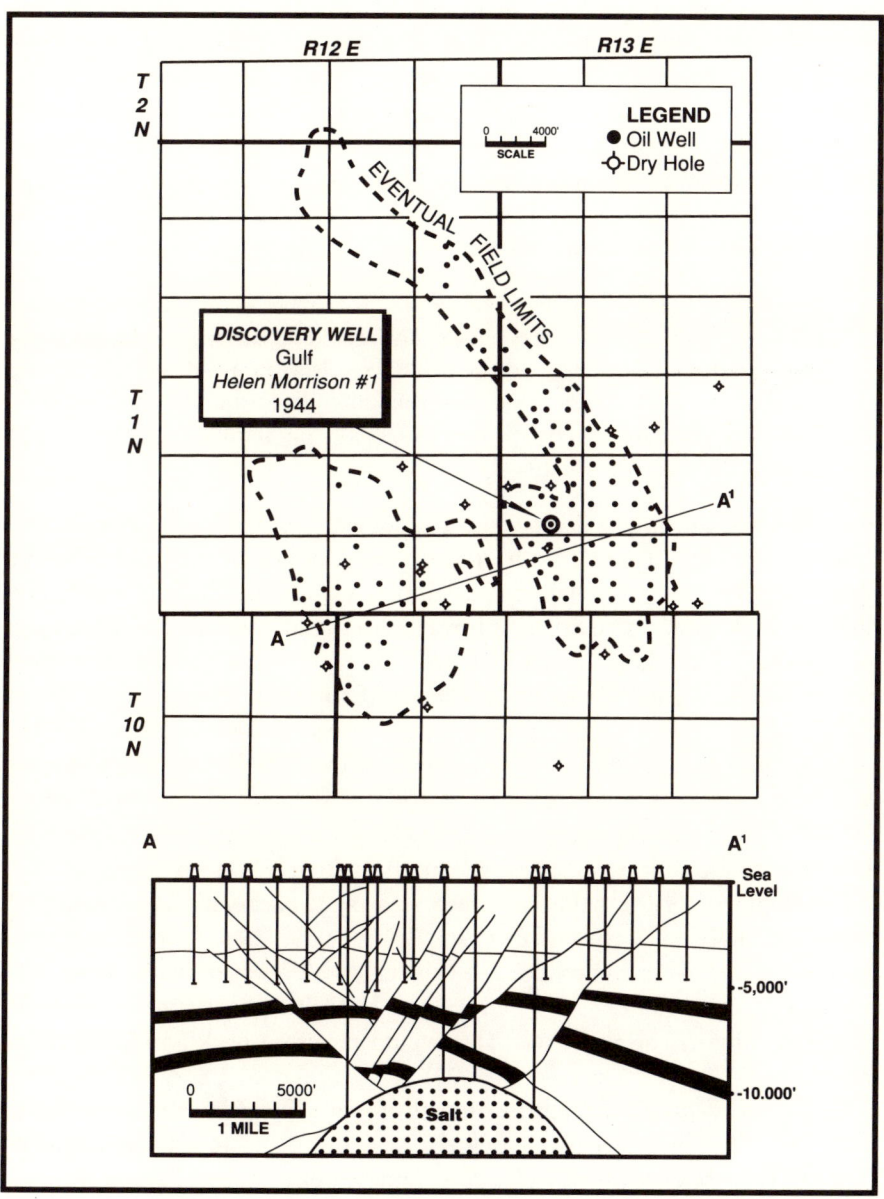

FIG. 11 Heidelberg Field, Wayne County, Mississippi, 1945

FIG. 11A Subsurface cross section interpreted from well control in 1960

into 20-acre drillsite tracts began in earnest, but the OPC insisted on 40-acre drillsites. Dixie comments, "If it were not for OPEC 40-acre spacing rule, there probably would be 10 rigs running in the Townsite by this date" (Dixie, March 30, 1944). With the big wells being found, many independents rushed in. Grady Vaughn from Dallas, Claude B. Hamill from Houston, B. J. Carraway of Dallas, and C. M. Beckett of Marshall, Texas and W. C. Proctor, Roger Lacy, P. G. Lake, the Frankel brothers, and J. B. White were among the newcomers:

> Heidelberg, Jasper County, Mississippi, is fast beginning to look like an oil field. Production derricks have been erected on the two producing wells and the derrick is still up on the Sun No. 1 Mack Lindsey, which gives a total of about 12 derricks that can be counted from the high hill at Gulf's No. 1 T. D. Lewis. The oil at Heidelberg was selling for 82 cents per barrel. (Dixie, May 18, 1944)

Tom McGlothlin, Gulf's geologist at Heidelberg, set up headquarters in the Pinehurst Hotel in Laurel. Each morning he telephoned the head office in Shreveport to report the latest information on all the drilling wells. His high-pitched voice carried easily through the thin walls of the Pinehurst guest rooms. Years later it was revealed that some lease speculators tipped heavily to be assigned the rooms adjoining McGlothlin's.

One of the independents, Graham and Lewis, drilled its No. 1 W. H. Husband unit to find 299 feet of saturated oil sand. They elected to perforate with 2,300 shots, at a cost of some $5,000. This paid off, however, when the well flowed a record for the field of 1,896 barrels of oil daily. As the field developed, new fault segments were found, including West Heidelberg.

Development of the field continued at a rapid pace with 104 producing oil wells completed by the end of 1945. At that time, more than four million barrels had already been produced in the field, which was only two years old. Eventually, more than 250 oil wells on the Heidelberg structure would yield over 175 million barrels of oil. This would prove to be Mississippi's third largest field.

Gulf's two decades of relentless searching for oil in Mississippi had been rewarded. The discovery at Heidelberg was considered of sufficient importance to be publicized worldwide. The January 1944 "pony" edition of *Time* magazine, a condensed wartime version, carried the story. In Australia, 22-year-old Julius W. (Judy) King, a lifetime resident of Heidelberg, who was serving in the army, learned of the discovery of oil in his hometown from the *Time* article.

The Collier's *Magazine Story*

On February 10, 1945, *Collier's*, one of the nation's leading magazines, printed an article on the Heidelberg oil boom, "'Ole Miss' Strikes It Rich," by Harry Henderson and Sam Shaw:

> Eleven fields are now in operation and wildcats are being drilled in twenty-five counties. The greatest drilling activity is near the little town of Heidelberg. . . . But the whole state is swarming with geologists, roustabouts, scouts, roughnecks, tool pushers, riggers, drillers, lease hounds, wildcatters, speculators, lawyers, tipsters and gypsters, and that crowd of fast operators who appear wherever money is made and lost hand over fist. . . .
>
> Lease hounds tract geophysical crews to try to dope out the area they are converging on, so they can jump into its center and grab some leases. The geophysical crews get up in the middle of the night and drive sixty miles in a circle to escape them. . . .
>
> The competition is frantic and fantastic not only for leases and royalties, but for everything from a hotel bed to a square meal.
>
> All the major oil companies and several scores of independents are drilling and exploring Mississippi's substrata. They have unleased a flood of money, millions in leases and drilling with royalties still to come in what has been one of the poorest sections of America.
>
> Southeastern Mississippi, which has become the focal point of the boom, has never known wealth. Here there were never plantations, but only small farms, worked by owners too poor to own slaves.
>
> Until thirty years ago, much of it was pine forest. Now most of this has been cut off, leaving scrubby, stumpy land almost too poor for cultivation, to be ravaged by soil erosion. . . .
>
> From the oilman's view, probably the most significant fact about the Mississippi discovery is that the search for oil has now moved eastward.
>
> The big change is in people's pocketbooks—money. For the first time in their lives they all have it.
>
> One of the men who will probably end up a millionaire as a result of the [Heidelberg] discovery of oil is chubby C. B. Burns who runs just about everything there is in Heidelberg to run, namely, a cotton gin, a general store, a cement agency and a trucking firm. Burns came to Heidelberg sixteen years ago fresh from the University of Alabama's school of commerce. . . . he managed to build up a profitable business and put the profits back into land. He now owns 700 acres and two good wells, and has an interest in four others. . . .
>
> The king-size good-luck story of Heidelberg is Harry Eddy, a big, slow-footed whittler with a downhill gait, whose cup overfloweth with ironic

amusement at his new wealth. He owns 1,805 acres in Wayne, Jones, Jasper and Clark Counties, with seven wells and more to come.

He was fired as a rural mail carrier four years ago, he says, because a local political clique wanted the job for a supporter.

He waved a corner "Ah'm goin to build me a nice five-story hotel. Hit'll be the biggest building in town. Ah doan' need it. Ah doan' want it. Ah'm not gonna run it. But Ah'm gonna build it and put my name on it in lights just to make em look at it every time they set foot outdoors."

"Now everybody is runnin' to meet me," he drawls "with get-rich-quick schemes, 'n' Ah tell 'em all: 'Brother, you're too late. Ah'm already rich. You need it more than me. Do it yourself and git rich, too.'"

Two young Mississippians named Evon Ford and John Clark from Smith County got in on twelve wells by loading a car with $10,000 cash and pulling farmers in off the street. Then while one offered a farmer cash for his royalty, the other phoned a companion in the courthouse who gave the man's title the once-over. If it looked clear, they bought it.

Mrs. [Effie] Husband accepted $100 from a strange white man for a third of her royalty. A few days later, her son Norman, to whom she had given some land, accepted $280 for a third of his royalties. Not until a sister got $5,000 for a similar share did they realize the value of the rights. (*Collier's Magazine*, February 10, 1945)

Others Join the Oil Crowd

Heidelberg attracted several new Mississippians to the oil business, including: D. W. Skelton, T. J. White, L. B. Porter, R. L. Windham, and Bob Shoemake. Skelton, an agriculture teacher at Heidelberg High School, bought royalty interest with White, a butcher from Meridian. White sold the farm he had inherited from his parents and put the money in oil, which proved to be a very wise move. Both were very successful during the Heidelberg boom and in other oil plays thereafter. Porter was a teacher at Heidelberg High School, later to become "Judge" Porter. Windham and Shoemake, both from Collins, became active in most of the oil plays in Mississippi after Heidelberg and became large mineral rights owners.

Mallalieu

Flush with the success of three consecutive discoveries (Brookhaven, Lake St. Johns, and Cranfield) The California Company started a well at Mallalieu Church. This structure had also been found by the company's first seismic line shot in 1937.

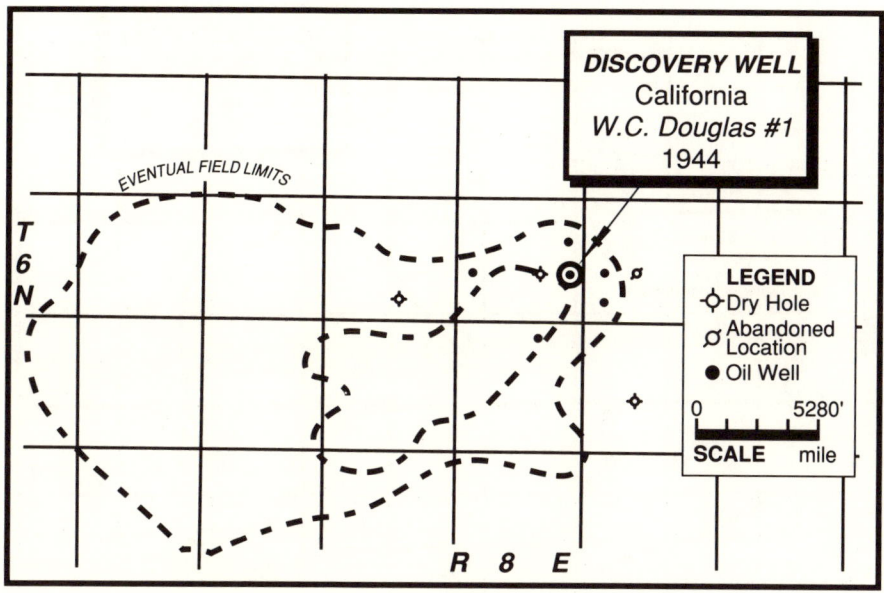

FIG. 12   Mallalieu Field, Lincoln County, Mississippi, 1945

On August 25, 1944, the company officials were elated when the No. 1 W. C. Douglas well flowed 374 barrels per day of 38-degree gravity oil from perforations at 10,520 to 10,560 feet in the Massive Sand of the Lower Tuscaloosa formation.

However, the sand conditions at Mallalieu were similar to those at Brookhaven, with a partially bald crest, so that follow-up drilling was somewhat erratic. The first offset to the discovery was dry with the sand shaled out. The second try in a different direction proved successful, but a third again found the sand almost shaled out and made a marginal well.

By the close of 1945 there were only three good wells and the importance of the field was very much in question. Further development after the war proved up the field which ultimately produced some 35 million barrels from over 100 wells.

Vaughey & Vaughey would be one of the luckier operators in the field after the war:

> When we were leasing what turned out to be the Mallalieu Field in Lincoln County, I received permission from the California Company to buy

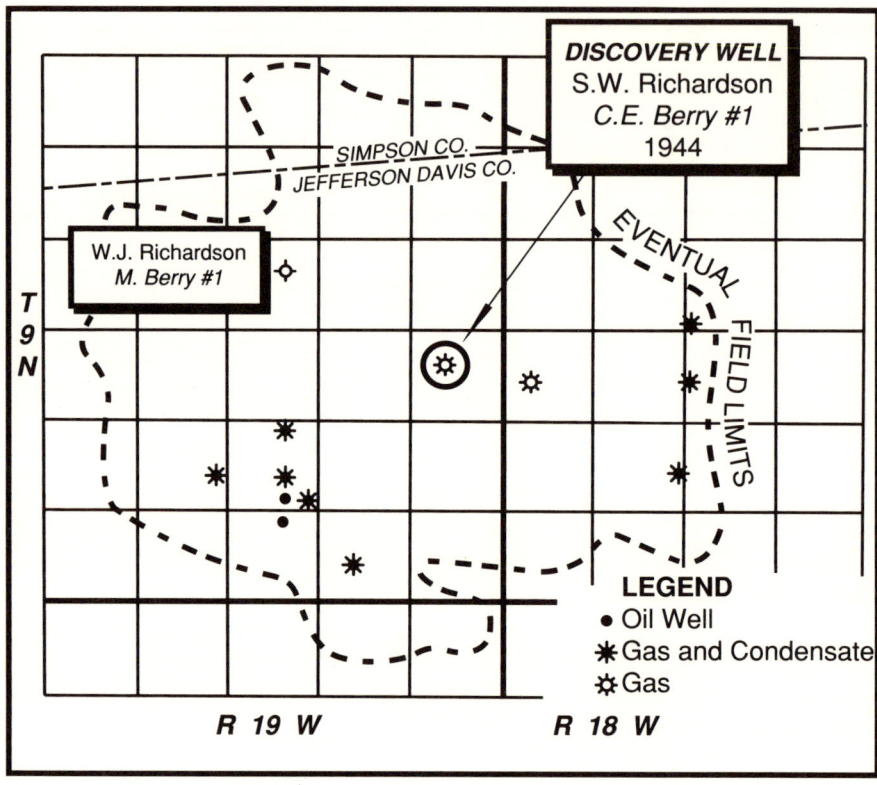

FIG. 13 Gwinville Field, 1945, became Mississippi's largest gas field

minerals for ourselves. Among the minerals we bought was one outside their block. After Mallalieu was pretty well developed, we thought this piece might have a chance of producing, we had the majority of the minerals anyway under it, and I leased the rest from people—landowners and people that had bought royalty—and drilled a well. We struck a terrific section of sand and this well also produced a million barrels of oil. California Company offsetted and got a dry hole, and we offsetted and got a dry hole. It was just a little one-well field of its own, very peculiar but lucky. (E. A. Vaughey, March 11, 1987)

Gwinville Gas Field

During the war years the largest gas field ever to be found in the southeastern states was discovered. This field, Gwinville, would eventually prove to have 10 times more gas reserves than the Jackson Gas Field.

Located in Jefferson Davis County, Gwinville is a massive deep-seated salt dome that uplifts beds over a 50-square-mile area. Gulf first leased this prospect in the early 1930s after doing pioneer gravity work.

By 1940 the structure still had not been drilled. With the leases nearing their expiration dates, Gulf farmed out to W. J. Richardson of Fort Worth, Texas, with the requirement to drill an 8500-foot well on the Maggie Berry tract. Richardson's well was dry, narrowly missing the field. Although the well seemed to be high regionally, Gulf elected to let its block fall apart. By 1944 at least ten companies and independents had acquired leases on the giant structure.

The next person to drill the structure was Sid W. Richardson, brother of W. J. Richardson. Sid worked out a nine-section "dry hole contribution deal" with the other companies in the area. By June 1944 he had drilled to a total depth of 7,628 feet in his No. 1 C. E. Berry well. A Schlumberger electric log was run on the well. Cores taken in the Eutaw section recovered sand with good odor and a taste of gas distillate. This was the first well in the state to find gas in the Eutaw or Upper Tuscaloosa sections. The well had 155 feet of productive gas sand. There was no question that a significant field had been found.

Surface location of the well measures 42.5 miles south of the city of Jackson. This was important to the Mississippi Power & Light Company, whose franchise to supply natural gas to the City of Jackson provided that if gas was discovered within a 40-mile radius of Jackson, the gas rates must be lowered to the original 30 cents per thousand, as was charged when the Jackson Gas Field was furnishing the supply (Dixie, July 13, 1944).

After reaching a total depth of 8,967 feet, seven-inch casing was run on the well. The well was completed from the "Gholar One" and "Gholar Two" sands of the Eutaw-Upper Tuscaloosa section, names local to this field.

Flow from the Gholar One Sand at around 7,800 feet was 3.5 million cubic feet per day, while flow from the Gholar Two section around 8,100 feet was 3.75 million cubic feet per day with a small amount of condensate (Dixie, July 27, 1944). In October 1944 Gulf extended Eutaw production 2.5 miles to the southwest with its No. 1 D. Lee Mullens.

Gulf drilled the Mullens well deeper and eventually completed it as an oil well in the Lower Tuscaloosa. Time would prove that oil reservoirs on the structure were minor, and Gwinville remained basically a gas field:

> The Superior Oil Company is rigging up one of the largest Rotary Rigs yet to move into Mississippi on their #1 Durr, Jeff Davis County, Mississippi, in the Gwinville Field. It is steam with five 250 H.P. boilers, three 20"

mud pumps and includes a standby draw works. A 176' derrick with
32' base is being used. (Dixie, October 26, 1944)

The Willie Durr well, located one mile to the east of the discovery, was
completed as the second well in the field. The Dixie Report of March 29,
1945 stated, "GWINVILLE MAY PROVE TO BE NATION'S MOST IMPORTANT DIS-
COVERY IN RECENT YEARS!"

> At present stage of development at Gwinville, an estimated 10,000
> proven acres of oil and/or gas production is indicated, and there is a good
> possibility that the productive area may go considerably higher.
> The Gwinville Field, Jeff Davis County, Mississippi, . . . has given
> evidence of being the state's largest producing area.

Gwinville would eventually prove to be the largest gas field in the southeastern states. By the end of 1945, 11 producing wells had been drilled, some
of them dual. Eventually, the field would have over 100 wells and produce
some 1.5 trillion cubic feet of gas and 10 million barrels of oil or condensate.

Gulf Finds Mississippi's Biggest—Baxterville

The largest oil field ever found in Mississippi was Baxterville in Lamar and
Marion counties. The Dixie Geological Report of August 5, 1943 indicates
the Gulf Refining Company had recently leased some 18,000 acres from
H. H. Bass of "paper shell pecan" fame. Consideration was $1.00 per acre,
and the lease provided that total acreage must be considered as a unit and
could not be selectively dropped. However, Gulf chose not to drill on this
lease with its first well.

In the late summer of 1944 Gulf began drilling its No. 1 C. V. Cooper.
By September, dry gas had been tested from the Eutaw formation and the
hole was being deepened to the Lower Tuscaloosa. Low-gravity oil was found
in the Massive Sands of the Lower Tuscaloosa section.

The well was completed on November 18, 1944, through perforations
from 8,690 feet to 8,744 feet, flowing at a rate of 609 barrels per day of 16.5-
degree gravity oil per day. The gas/oil ratio was 238 to 1. Although the well
tested at very good rates, the low gravity nature of the oil commanded a
reduced price, a disappointment to Gulf. Even so, the very thick pay sands
convinced made most oil men that a sizable reserve had been discovered.

The Superior Oil Company drilled the second well in the field, moving
in a large steam rig with a 176-foot derrick. The company drilled its No. 1
R. Batson 1.5 miles southeast of the discovery well, and proceeded to dually

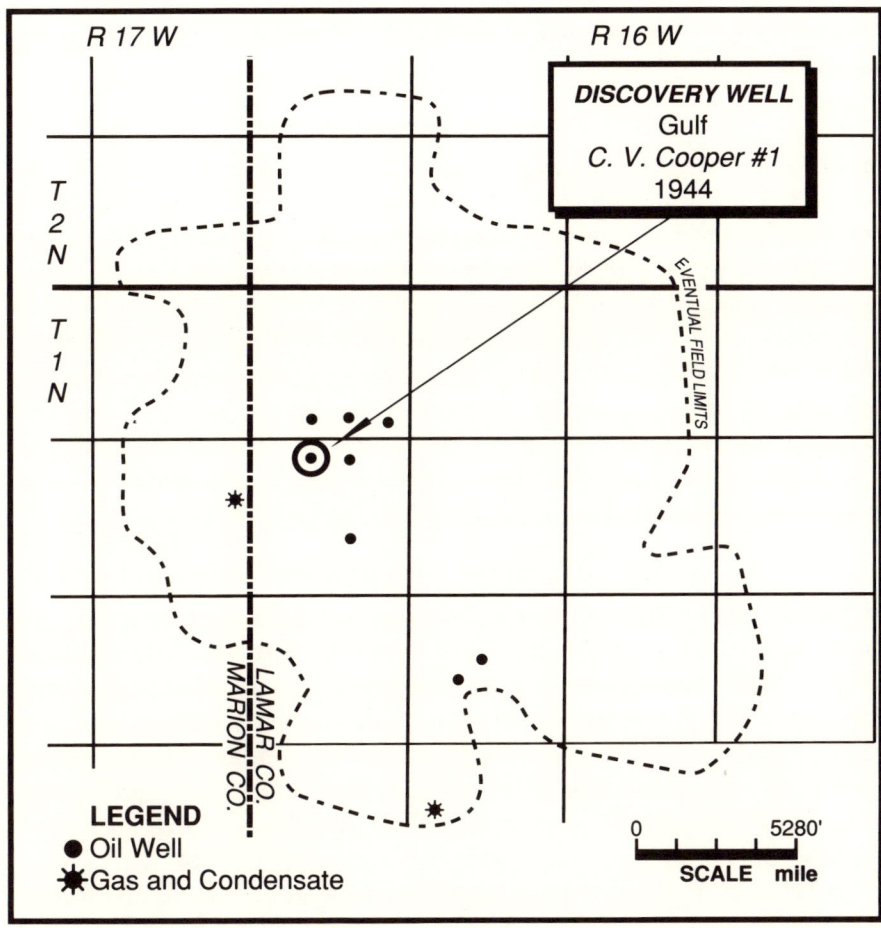

FIG. 14   Baxterville Field, 1945, eventually Mississippi's largest oil field

complete the well in both the Eutaw Upper Tuscaloosa gas zone at 7,800 feet and the Massive Tuscaloosa Sand at 8,800 feet. (This was accomplished by running two strings of tubing, separating the two zones with a dual packer.) The upper zone flowed 3 million cubic feet of gas per day while the lower zone flowed 582 barrels of oil per day; the well was officially completed on March 5, 1945.

Gulf went on to drill the J. M. Andrews well. Soon Sun and Texaco were drilling on the structure, and by the close of the year, 16 wells were producing. The field was slowly developed over a number of years and redrilled in the 1970s. More than 250 wells had produced almost 250 million barrels of oil and over 400 billion cubic feet of gas by 1990.

Kirby Petroleum Discovers Langsdale

Drilling in Alabama had established an east-west trending fault line, which became known as the Gilbertown Fault. This break in the formations ran in a westwardly direction across the state line into Wayne County, Mississippi.

In September 1944 Kirby Petroleum Company permitted a well on the Long Bell Lumber Company lands along this fault. The well was completed at a depth of some 3,700 feet, pumping 145 barrels per day of 18-degree gravity oil on January 18, 1945. The company continued to develop the field until 16 producers had been completed by the end of 1945. Hunt Oil Company extended its six-inch gathering system from Gilbertown to the Langsdale Field to take the oil. Eventually, some 40 wells in the field would produce approximately six million barrels during their lifetime.

Pure Oil Succeeds at Carthage Point

During December 1942 the Pure Oil Company, from its Jackson office, began to drill a well in the Mississippi River floodplain north of Natchez. In early 1943 the company's No. 1 J. M. McDowell well was below 10,000 feet. An open hole drillstem test flowed gas from the Lower Tuscaloosa Sand at a rate of 7 million cubic feet per day plus 15 barrels per hour of 47.5-degree gravity oil, but with some saltwater. Pipe was set, and the production man, Frank Manning, tried unsuccessfully to complete the well for several months.

On February 13, 1943, through perforations at 10,281 to 10,285 feet in the Massive Sand of the Lower Tuscaloosa, the well flowed for 24 hours at a rate of 100 barrels of 40-degree gravity oil per day and 1,283 MCF of gas per day. However, the well died after 10 days of production, and attempts to revive it failed.

The operation was plagued by floodwaters from the Mississippi River along with an abnormal number of downhole problems.

Pure had every intention of drilling a second well immediately, but due to many delays the second well was not started until the end of 1944. It was completed on January 8, 1945, flowing 91 barrels per day of 54-degree gravity oil plus 2 million cubic feet of gas per day, and thus received credit as the discovery well.

A three-inch gas line was laid to the old Interstate Natural Gas Company line that transported gas from the Monroe Field to the Standard Refinery at Baton Rouge.

At the end of 1945, Humble Oil & Refining Company was drilling a second well, the No. 1 C. L. S. McKittrick, half a mile east of the discovery well. Carthage Point proved to be a minor field with some 15 wells that would ultimately produce some 50 billion cubic feet of gas and 5 million barrels of oil.

Soso—Gulf Scores Again

As 1944 came to a close, Gulf was drilling its No. 1 Edwards-Bailey well near Soso in Jasper County, Mississippi. The company chalked up another success: "The Gulf Refining Company has hit the 'jack-pot' again, this time with a potential discovery on their No. 1 Edwards-Bailey . . . in the SoSo area" (Dixie, December 21, 1944). Initial cores in the Upper Eutaw (Stanley Sand) appeared to be wet, but the underlying Christmas Sand was found productive. Later tests would prove that both were productive. The well was completed in March 1945, from the Christmas sand of the Eutaw formation at a depth of 6,600 feet. The well flowed five million cubic feet of gas per day with 156 barrels of condensate.

Soso was a large shallow gas field that ultimately produced 118 billion cubic feet of gas and 2.5 million barrels of condensate. However, no additional wells were drilled during 1945. Of more significance, a large oil field was discovered at greater depths on the Soso structure in the early 1950s.

Humble at Fayette

Humble made the next discovery in Mississippi with the No. 1 M. R. Smith well in Jefferson County. The well found oil and gas condensate in the Lower Tuscaloosa similar to that at Cranfield. By this time, the royalty buyers had more confidence in this type of production and royalty sold for up to $300.00 per acre.

This field, Mississippi's fourth new field discovery in 1945, was named the Fayette Field. The well flowed 229 barrels of 37-degree gravity oil per day from the Lower Tuscaloosa at 9,700 feet. No additional wells were drilled by the close of 1945. This fairly small field, ultimately producing some 5 million barrels of oil and 26 billion cubic feet of gas.

Humble's Second Success—Hub

Humble was more successful in Marion County, near Columbia. The company discovered the Hub Field with completion of its Evans No. 1 well on

February 25, 1945, at 9,100 feet in the Lower Tuscaloosa. The well flowed gas at the rate of 1,420,000 cubic feet per day and 120 barrels of 43-degree gravity oil. The California Company then moved to the large W. E. Walker lease approximately a half a mile to the northeast. The Walker well was also successful. Humble drilled its No. 1 A. E. Ball to establish the third producer in the field.

The year 1945 ended with three gas wells completed in the Hub Field. Hub proved to be a significant gas field, eventually producing over 350 billion cubic feet of gas and eight million barrels of oil.

CHAPTER 37

Near Misses

In addition to the oil discoveries during the war years, a number of wildcat wells found very favorable conditions. But these wells were "high and dry" for no apparent reasons. In some cases, later drilling found good fields on the same structures.

One such structure was in Jones County, Mississippi, where Union Sulphur drilled two wells, both dry. These wells proved to be immediate downdip from the Sandersville Field, which would be discovered by Central Oil Company during the 1950s. Merrill Harris worked on these wells, as a consultant for Union Sulphur.

On the southern end of the same huge, deep-seated salt dome, Gulf drilled its Sandersville Block, but gave up after three dry holes. Later these would prove to be shallow wells over the Pool Creek Oil Field to be found at a greater depth in the 1950s by Zack Brooks.

Southern Natural Gas entered the picture and drilled the Satartia structure, which had been surface-mapped at the same time as Tinsley by the crews working with Fred Mellen. The No. 1 Gammill well found shows in the Eutaw and attempted a completion that was unsuccessful. Hughes & Hughes discovered a small oil field, Satartia, at a greater depth in the 1960s.

Sinclair had been very unlucky in Mississippi in spite of intense exploration activity over the years. In January 1943 Sinclair abandoned its No. 1 Warsaw Lumber Company well in Jones County, which was apparently on a large structure. The Dixie Report cited this as "1942's most disappointing wildcat" (Dixie, February 21, 1943). A second well was drilled, which was also dry: "the Tuckers Crossing Prospect . . . was first turned to Sinclair in 1939 by Alfred Foote and Urban Hughes and was one of the pre-Tinsley prospects of Mississippi" (Dixie, June 8, 1944). Both of these wells were "high and dry" but set up the structure that produced in the 1960s as the Wausau Field, found by Chismor, a partnership between A. F. Chisholm and Jimmy Morgan.

Gulf drilled a wildcat near Taylorsville in Smith County, Mississippi, in early 1945. Its No. 1 E. S. Gambrell well tested gas from the Eutaw but in noncommercial quantities. This structure later proved to be on the edge of the Summerland Field to be found by Triad Oil & Gas in the early 1960s.

Late in the year 1945, Humble Oil & Refining Company drilled the No. 1 Waer in Greene County, to find Sand Hill, one of the largest salt structures in Mississippi. Production was eventually found on this structure, but in spite of its size, it had no significant reserves.

In 1943 The California Company obtained a shooting option from Greene Lumber Company on 12,000 acres in Smith County. Seismic studies indicated a structure, which later proved to be the Raleigh Field, though it was not drilled until the 1950s. Near the end of 1945 Humble Oil & Refining Company made a number of locations, one of which would help set up the West Yellow Creek Oil Field in Wayne County, to be drilled later by independents.

On the Wiggins Arch, Harry Morgan and S. P. Borden continued to drill wildcats in Stone County, on Dancler Lumber Company's land. They missed Maxie and Pistol Ridge, which were found in the early 1950s. In Rankin County, east of Jackson, the Phillips Lumber Company began drilling on the large Denkmann Lumber Company leases. The well tested the Cotton Valley but was dry. Five years later, Phillips would find a huge reserve of almost pure carbon dioxide in the Smackover on the Denkmann lands some 2,000 feet deeper.

A deep well was drilled at Tinsley near the end of 1945. Union Producing Company drilled the No. 21 Jennie Stevens to a depth of 11,625 feet in the Smackover lime. The well recovered gas-cut muddy salt water from the Smackover zone. A "sour gas" well was drilled two miles away on the structure several years later, but was never produced. These "heartbreakers" tempered enthusiasm for drilling in Mississippi in spite of successes on other structures. Around Natchez many shows were being encountered in Wilcox Sands, apparently from "stratigraphic traps."

With the successful wells in Louisiana and at Cranfield, interest in the Wilcox continually expanded. In late 1944 Humble released information on a well that had made oil in noncommercial amounts from the Wilcox in Adams County, the No. 1-A Stowers. In February 1945 Humble also reported that its No. 1 Smith well in Jefferson County, Mississippi, had encountered gas and oil shows in the Wilcox. There was no successful completion.

In mid-1945 Charles H. Osborne drilled the No. 1 Armstrong-Ellislie in Adams County and encountered good shows of both oil and gas in the Wilcox that "correspond roughly to the zone of non-commercial oil shows at Cranfield." The Wilcox had been partially successful at Cranfield, but "cementing off" water sands was very hazardous:

> "The fifth Wilcox producer is assured at Cranfield with the California Company No. 1 L. G. Ratcliff Unit." (Dixie, June 21, 1945)

Later in 1945, Dixie reported:

> In October, 1945, Roeser & Pendleton-Sohio Petroleum Company No. 1 Baker-Maier recovered oil shows in the Wilcox which indicates possibly a new field discovery. . . .
> This is the third test to be drilled on the prospect. The first was a dry Wilcox test, Elliott No. 1 Ogden; the second a deep test, J. P. Evans No. 1 Parker which was drilled to the top of the Comanchean with no massive sand present. (October 25, 1945)

Thus 1945 ended with the Wilcox Trend showing promise, but problems with completions had not been satisfactorily solved. Only Cranfield had commercial Wilcox production. The numerous shows indicated probable Wilcox fields and most would prove to be in or near Wilcox oil fields. Few people guessed that almost 250 Wilcox fields would be found in the last half of the century.

Salt Domes

During the 1942–45 period, 18 additional salt domes were found in Mississippi. Nine were found by Gulf, five by Freeport-Sulphur Company, one by Humble, one by Sun, and one by C. H. Osmond. Piercement salt domes had been very disappointing in Mississippi to this time, even though they had provided traps for excellent fields in Louisiana and Texas. Geophysical information had warned that these salt domes were of the piercement type, but the hope was still alive that production could be obtained over these domes similar to that in Spindletop.

Apparently vindicating these hopes, the Magnolia Petroleum Company No. 1 Pidgeon was drilled on the King's Dome in Warren County near Vicksburg in early 1942. A Sparta Gas Sand blew out at a depth of less than 1,200 feet and caught fire, causing considerable damage to the rig. This was the fourth core test on the dome and was the highest well structurally. A heavy oil show was also encountered in the top of the Wilcox. Even though this was a noncommercial well, it encouraged people to try other piercement salt domes.

On January 6, 1943, H. C. Peterson, a geologist with Freeport-Sulphur Company of New Orleans, addressed the Mississippi Geological Society on the geology of salt dome cap rock associated with sulfur deposits:

> [He] stated that the total cumulative Salt Dome Sulphur production to date is approximately 55 million tons. Therefore, there was still hope that some of Mississippi's salt domes would find a sulphur cap that would be very valuable. The Ruth and D'Lo Dome found in 1942, Sardis Church, Lampton, Leedo, Byrd, New Home, Prentiss, Carson, Monticello, and Moselle, were found in 1943. (Dixie, May 4, 1944)

> In early 1944, Freeport-Sulphur found the Bruinsburg Salt Dome. In July, 1944, Sun Oil Company decided to drill another test well on the Bruins-burg Dome. Their No. 1 W. R. Hammett Well blew out on July 16th, at a total depth of 1837 feet, probably in the same Sparta sand as the King Dome Well. The well was completed as a gas well of low volume, but did not prove to be commercial. However, Sun was given credit with discovery of the Bruinsburg Field. (Dixie, July 20, 1944)

Three additional domes were discovered in 1944: the Allen Dome, the Richton Dome, and the Arm Dome. In 1945 the Hervey and Galloway domes were discovered.

By the close of 1945, 27 salt domes had been found in Mississippi, all with no commercial success. Salt domes became unpopular as an exploration target after this time, even though many more were found over the years.

CHAPTER 38

The Oil Fraternity of the War Years

In 1942–45 some of the people and companies active in the Mississippi oil business were J. H. Hewgley (Drilling Co.), Roy Lee (with Hassie Hunt), A. A. Holston (Stanolind), C. F. Kimball and J. B. Wheeler (Sinclair), Lester and Witcher (abstractors), and H. L. Spyres and Charles Buck (Skelly). Individuals included Reese Oliver, R. L. Fisher, Pres Cochrane, Harry Elliott, Paul Thomas, Brame Womack, T. R. Stubblefield, Collie Falk, Carl Grubb, Claud B. Hammill, J. B. White, W. B. Hinton, T. F. Hodge, Asa R. (Rags) Matthews, Grady Vaughn, and Dorris Ballew. Roy T. Whiteman, formerly in the land department of The California Company, resigned in the summer of 1943 to become associated with Vaughey & Vaughey in Jackson. Frederic F. Mellen became district geologist with the British-American Oil Producing Company for a brief period in 1943, but soon resigned to become a consultant in partnership with Mike Monsour.

J. Willis Hughes

Not mentioned in the Dixie Reports was J. Willis Hughes, who arrived in Mississippi in February 1943. He was transferred to Mississippi by Atlantic Refining Company and was in charge of four seismic crews working east of the Mississippi River.

Hughes was born in Oklahoma Indian Territory in 1905, near Ardmore. After high school he attended Rice University and Texas University. The Depression caused him to end his studies before receiving a degree. As early as the 1930s he became involved in the oil business by participating in several wells in Texas. When these wells did not prove profitable, Willis accepted a job with GSI, a seismic company then doing work for Atlantic Refining Company. Willis worked for Atlantic through the war years, but became a well-known independent in Mississippi after the war.

Homer Lynn

In November 1940 Magnolia Petroleum Company transferred Homer Lynn to Mississippi, to be stationed in Meridian. Lynn, a native of Oklahoma, had been a scout for Magnolia since 1934. Lynn began his new job as land-man, working with Magnolia's district landman in Jackson, Red Selby, who later resigned to become an independent lease broker.

As the war progressed, Lynn was promoted to the company's Dallas office in the Magnolia's famous building with Pegasus, the flying red horse, promi-nent on its roof. This was the tallest building in the state of Texas. During this period his salary was increased from $150.00 per month to $500.00, very good pay at that time. However, Lynn missed Mississippi. When Selby offered him $5,000 to quit Magnolia and move back to Jackson to work with him, he accepted the offer. It was necessary for the War Manpower Com-mission to approve this move. The purchasing department of Magnolia helped him to obtain a car, a very difficult feat under wartime regulations. In 1944, Homer Lynn became a permanent resident of Jackson. He has been a highly respected lease broker in the southeastern states since that time, for over 45 years (Lynn, 1990).

J. E. Stack, Sr.

J. E. Stack, Sr., a lay preacher and oil man from Shreveport, became a permanent resident of Mississippi in 1943. Stack had been in Mississippi during the short-lived Topton boom in 1929, and had become enamored with a lady in Meridian, which he chose as his headquarters. The Dixie Report of July 22, 1943, reported that Stack had "taken out a permit to drill the #1 Flintkote Company."

His son, J. E. (Jack) Stack, Jr., was serving in the navy at the time, but would return to Mississippi after the war to become an independent oil operator.

Companies' Activities

In regard to the Humble Oil & Refining Company, it is interesting to note that they are now in number one place in total acres under lease in the state of Mississippi. The Gulf Refining Company was the leader, but during the year 1941 they dropped considerable acreage. (Dixie, March 12, 1942)

The Sinclair Wyoming Oil Company has been tops from the wildcatting standpoint up until the year of 1942, but seems to have a reduced budget for this year. (Dixie, May 28, 1942)

The Louisiana Land and Exploration Company seems to be concentrating on minerals and Royalties instead of leases. They have been buying considerable minerals in the River counties in Mississippi, especially Adams County. (Dixie, June 11, 1942)

The list of new comers and comebackers to Mississippi continues to grow. The Shell Oil Company probably tops the list with personnel. They have rented six rooms in the Tower Building in Jackson. (Dixie, September 9, 1943)

The question has been raised as to why Shell Oil Company was moving a large number of personnel to Mississippi not having anymore holdings than they do. As a matter of fact, about all they have on record is the block in Neshoba County, which they are now drilling with the Southern Natural Gas Company. (Dixie, November 4, 1943)

Actually, Shell kept a large geological staff in Mississippi for many years before beginning significant exploration in the 1950s.

Other companies opening or enlarging offices in Jackson during the war were Phillips, Navarro, Arkansas Fuel Oil, British-American, Stanolind, Continental, Skelly, Superior Oil Company of California, Sohio, McAlister Fuel Oil, Pan-American, and Shamrock Oil and Gas.

The Scouts

The 1944 officers of the scouting association were: Gilbert Talley, Lion Oil Refining, president, Clyde Bale, Sinclair-Wyoming, vice president, and Ray Stevens, Shell Oil Company, secretary-treasurer.

Charlie Seedle, Arkansas Fuel Oil Company, replaced Bale as vice president when Bale was inducted into the navy.

New scouts in Jackson during 1943–1944: J. L. Blackwell, The Texas Company; Newton Burnett, Sun Oil Company; C. B. Lisman, Gulf Refining Company; C. D. Lambert, Continental Oil Company; J. H. McCready, Kingwood Oil Company; Dean Rogers, Stanolind Oil and Gas Company; Vent B. Speaker, Southern Natural Gas Company; C. W. "Jack" Stimson, Phillips Petroleum Company; and Claude Teel, Skelly Oil Company (Dixie, March 30, 1944).

The Mississippi Oil Scouts will be host to the first post-war National Oil Scouts Association Convention, which will be held in Jackson on May 23, 24, and 25, 1946. The Heidelberg Hotel is to be headquarters. (Dixie, November 8, 1945)

The 1945 Mississippi Oil Scouts officers were: E. K. Scott, Superior Oil Company, president; W. E. Keller, Lion Oil Refining, vice president; and R. E. Stevens, Shell Oil Company secretary-treasurer and editor.

Other scout-check members who would become longtime Mississippians were: B. G. Bushmier, Arkansas Fuel; H. K. Kent, Texas Company; C. H. Mayeaux, Continental Oil Company; V. W. Michael, Amerada Hess Oil Company; and J. F. McWilliams, Union Producing Company.

Landmen

The Mississippi Association of Petroleum Landmen was formed on June 7, 1944, the second landmen's association to be formed in the United States. The first had been formed in the tri-state area of Illinois, Indiana, and Ohio in 1939. However, when oil was discovered at Tinsley in September 1939, the early founders of the Tri-State Landmen's Association relocated to Jackson.

The Mississippi Association of Petroleum Landmen chose W. Brantley Jackson as its first president. C. H. "Red" Fidler, Bob Hart, and "Dutch" Miesse became charter members. The Mississippi organization was formed 10 years ahead of the national association, which did not come into existence until 1954. The Mississippi association is recognized as the "oldest continuously active landmen's association" by the American Association of Petroleum Landman (Joe Mayeaux, 1991).

Dixie Adds Clyde Alexander

With the expanding oil business in the southeastern states, the Dixie Geological Service also expanded. It recruited Clyde Alexander, who would spend the rest of his career in Mississippi.

Alexander first appeared on the Mississippi scene in 1942:

> The Mississippi Geological Society held a regular dinner meeting at the Edwards Hotel, Wednesday night, May 20. Mr. Clyde Alexander with the Phillips Petroleum Company at Shreveport, Louisiana, participated in a very interesting discussion of the problems of drilling, testing and completion of Wilcox wells. (Dixie, May 21, 1942)

Two years later, the following announcement was made in the Dixie Report:

> WE TAKE PRIDE IN ANNOUNCING THAT CLYDE W. ALEXANDER WILL BE
> ASSOCIATED WITH US AS A PARTNER IN THE DIXIE GEOLOGICAL SERVICE, AS
> OF APRIL 1, 1944!
>
> Clyde has been District Geologist for Phillips Petroleum Company in
> Mississippi. He was first employed by Phillips in 1935, and 'sat on' his first
> well in the Rodessa Field, Caddo Parish, Louisiana.
>
> We consider Clyde to be one of the outstanding Lithologists in this part
> of the Country, and his experience and ability as a Petroleum Geologist
> are such that we think we have scooped the industry.
>
> We three met at Fort Worth, Texas, in the 1920's, when Bud [Norman]
> and I [Buzz Morgan] were in School and Clyde played on the sand lot
> teams. Clyde was born one block from the T.C.U. campus and practically
> grew up with the students. His father, Charles Ivan Alexander, Sr., was
> Professor of Mathematics until his death in the fall of 1919. Clyde enrolled
> at T.C.U. as a freshman in fall of 1930, one year after Bud and I were
> gone, so we did not get to attend to his freshman upbringing. (Clyde said
> he didn't regret it!)
>
> Clyde and I went through the Rodessa Boom, then Schuler and Mag-
> nolia, Arkansas, and both of us ended up in Mississippi. It should be
> permissible for we three to go through Mississippi together.
>
> And Now! Us Frogs don't want no static outa you Sooners, Cajuns,
> Steers or Farmers!

In later years, Alexander and Norman, in partnership, each probably did
the geological and engineering work on more wells than any other two geologists in the southeastern states.

Other bits of news reported during the war years:

> We have recently received a V-mail letter from Lt. Sam Vauclain III, who
> is stationed in New Guinea, acknowledging receipt of a recent Dixie Geo-
> logical Service report. At present Sam is our most distant reader, but all
> of us who know him as Sun Geological scout in Jackson, would be glad to
> resume a local address. (Dixie, June 15, 1944)

The Engineers

With the many new fields being found in Mississippi, more petroleum engineers were needed to complete and produce the hundreds of wells being
drilled. They could use the reservoir information being collected on each
company's wells to great advantage in understanding the characteristics of
Mississippi's fields.

It was decided to organize an "engineering committee" to accumulate data on Mississippi oil and gas fields in 1944. This information could be furnished to the state Oil and Gas Board for the purpose of achieving the most efficient operating practices (Dixie, December 21, 1944).

The Mississippi Oil and Gas Engineering Committee was formed on January 12, 1945, with officers: D. V. Carter, Magnolia, general chairman; L. H. Moore, Gulf, vice chairman; Bud Norman and Buzz Morgan, Dixie, secretary-treasurer; and J. D. Davis, Union Producing, chairman-executive committee.

Members of the executive committee were: A. S. Rhea (Sun); L. H. Moore (Gulf); J. J. Bresnham (Magnolia); J. W. Gill (Hunt); Paul F. Barnhart (Frankel Bros.); and J. S. Boldrick (Humble). As a full-time director of the engineering committee, the organization hired Alec M. Crowell, who had recently resigned as director of production of the Arkansas Oil and Gas Commission (Dixie, January 18, 1945).

At a meeting on May 5, 1945, a major project was undertaken. The engineering committee would determine the "gas-oil ratio and reservoir pressure" for the principal fields in the state. To accomplish this, a committee was appointed: W. W. Ramseur (Gulf), chairman, with members C. M. Roberts (Union Producing), John Suman (Humble), John Pettigrove (Phillips), R. M. Hurt (Carter), and J. J. Bresenham (Magnolia). Field engineers hired to do the work were John K. Wright, Jr., and Carl Gruberman (Dixie, June 7, 1945). In later years the engineering committee would become the Society of Petroleum Engineers (SPE).

Mississippi Geological Society

The intensive drilling program in Mississippi during the war years kept many geologists in the field much of the time. Nevertheless, the Mississippi Geological Society continued to function.

Some of the speakers during the 1942 to 1945 period were George Schneider (Texas Company), Roy T. Hazzard (Gulf), Dr. Fish (Louisiana State University), Jules Braunstein (Shell), Joe Wheeler, Ed Cram, and Watson Monroe.

Officers for the 1942–43 season were Dave Harrell (Carter), president; L. R. McFarland (Magnolia), vice president; and K. K. Spooner (Atlantic), secretary-treasurer.

For the 1944–45 season L. R. McFarland was president; J. B. Storey (Union), vice president; and Fred Mellen (British America), secretary-treasurer. When the war ended in 1945, an influx of geologists returned from military service.

Mid-Continent Oil & Gas Association

The oil business established a very high profile in Mississippi with the proliferation of oil and gas wells. New wealth was being poured into the state. It was Mississippi's fastest growing business during the 1940s.

The state legislature gave top priority to oil and gas matters. Bills were being considered from all sides to tax and control the industry. Most of the legislators had little understanding of the complex business.

It became apparent to leaders in the petroleum sector that a unified front must be presented to the lawmakers.

> On Friday night, October 20, a state-wide meeting of persons interested in the oil industry was held at the Walthall Hotel in Jackson, Mississippi. The attendance was well over 100, and at this meeting it was decided to form a Mississippi-Alabama Oil and Gas Association. An organization committee was named and the meeting was adjourned until Friday night October 27, at which time a Board of Directors will be elected and it will be decided whether to affiliate with the Mid-Continent Oil and Gas Association or to continue as an independent organization. (Dixie, October 26, 1944)

> On Friday night, October 28, the organization meeting of the Mississippi-Alabama Oil & Gas Association was held, and by vote of those in attendance decision was made to accept the invitation of affiliation with Mid-continent Oil & Gas Association: to be known as Mississippi-Alabama Division of that parent organization.

> The board of directors was selected, pending acceptance by those not present, and the following officers were then elected: C. L. Morgan, president; W. M. Vaughey, vice president; and Buford Yerger, temporary secretary. (Dixie, November 2, 1944)

Edward D. Kenna was hired to be the organization's first chairman. The Mid-Continent became the mouthpiece of the oil industry in Mississippi and Alabama. Since its beginning, the organization has been a major influence in shaping oil and gas legislation to the benefit of the people, the state, and the industry.

CHAPTER 39

Mississippi's Wartime Success

Mississippi had done its part during World War II. New large fields had been found and monthly oil production in the state consistently exceeded its quota.

As the war ended, Mississippi still had not passed a conservation bill or established statewide spacing rule.

> A conservation measure had been presented, and defeated in 1942 and 1944, Mississippi legislative sessions: the program is outlined to be patterned after the Arkansas laws which have operated so successfully.
>
> We are of the opinion that failure of the previous bill has resulted from inadequate education of the general public with regard to the industry new to the state. In order that they may make their own decision, every effort should be made to familiarize Mississippians with comparative results of controlled and uncontrolled production methods as demonstrated in older producing areas.
>
> Mississippi is in a particularly favorable position in this respect, because it is able to profit by results obtained through many years of other states under conditions similar to those we face at present and are in the future to encounter again and again. (Buzz Morgan, Dixie, December 21, 1944)

The Mississippi Oil and Gas Board began a series of public hearings in March 1945, drawing great crowds as it discussed conservation measures. The independents wanted 10-acre oil spacing, except a few such as the Vaugheys and Buzz Morgan. Companies wanted 40-acre oil spacing. On May 14, 1945, however, the Oil and Gas Board issued Order No. 3-45, which in

FIG. 15   Oil and gas fields of Mississippi and Alabama, 1945

effect said that the state would continue to be governed by the rules of the PAW (Petroleum Administration of the War). This convenient side stepping of the issue only lasted to the end of the war.

The Mississippi Oil and Gas Board received official notice from the PAW that its orders would become null and void on September 1, 1945. This forced the board to act. It finally adopted statewide spacing rules, which were very similar to the PAW regulations, providing that oil wells must be located on 40-acre spacing and gas wells on 320-acres, with minimum distances from lease lines. However, these rules did not have the solid backing of law and could be waived to suit political expediency.

A new federal agency appeared on the Mississippi scene, the Federal Power Commission (FPC), which was investigating natural gas. The governor appointed a steering committee of oil men to assist the state appearing before the FPC. These were Buzz Morgan, Bill Vaughey, Ed D. Kenna, Leon

Tyrone, E. O. Spencer, Ben Cameron, D. V. Carter, L. H. Moore, David Gray, H. M. Morris, and Alec Crowsell. No one realized the magnitude of disaster that would befall the natural gas industry in the 1950s as a result of FPC control

Summary of Mississippi's Oil Business 1942–45

A tabulation in the National Oil Scout and Landman's Association Yearbook showed that between January 1, 1942, and December 31, 1945, a total of 344 oil wells, 16 gas wells, and 372 dry holes were drilled in Mississippi, or a total of 732 wells, during the war years. Some 5 million acres of land were under lease for oil and gas at the beginning of 1942. This increased to over 8 million acres by the end of 1945. Humble was the largest leaseholder.

Producing fields were listed as Cranfield, Carthage Point, Bruinsburg, Langsdale, Quitman, Jackson, Heidelberg, Soso, Fayette, Gwinville, Baxterville, Brookhaven, Mallalieu, Pickens, Flora, Hub, Amory, Carey, Eucutta, and Tinsley. Amory had been abandoned.

CHAPTER 40

Alabama, 1942–1945

Despite the large number of wells being drilled in Mississippi, Alabama received scant attention during most of the 1942–45 period. Alabama had the distinction of passing legislation to establish oil and gas laws before producing a barrel of oil. In 1940 the Alabama legislature set up an Oil and Gas Board consisting of the governor, the attorney general, the director of conservation, and the state geologist. A drilling permit costing $25.00 was required, as was a bond to ensure compliance with the provisions of the law. A two percent tax was to be levied on all oil and gas produced and sold. One interesting clause in the law was this:

> Nothing in this act shall be construed as authorizing the Board to limit the production of any well or pool or field until such time as the total average daily production of the State shall exceed 50,000 barrels per day for a period of 90 consecutive days.

In 1945, Alabama cooperated with the federal government to pass a set of oil and gas conservation laws along the lines recommended by OPC including 40-acre spacing. The Oil and Gas Board had been appointed and its rules and regulations were being administered by 1946.

Only minor drilling projects were reported in the early years of the war. The scout reports indicate that R. N. Ranger of Houston, Texas, made a deal to drill several wells on the Scotch Lumber Company land in Clarke County, Alabama. He also obtained a farmout on the Vaughey & Vaughey block in the same county, with the obligation to drill.

After Ranger's well on the Vaughey block was dry, James L. Duffy took a second farmout from the Vaughey brothers in 1943, and sold interest to Atlantic, Humble, Union Producing, Sun, and others for a 5,500 foot test. Superior drilled a well in the Black Warrior Basin, Winston County, which reported a small gas flow from the Bangor Lime.

Gilbertown—Alabama's First Oil Production

In late 1943 H. L. Hunt made a location for his No. 1 Robert T. Land near the Alabama line in Clarke County, Mississippi. The block had been assembled by Harry Elliott. Even though the well encountered oil shows in the Eutaw at approximately 3700 feet, the well was abandoned as a dry hole. Not discouraged, Hunt moved further east into Choctaw County, Alabama, and drilled his No. 1 Jackson well:

> H. L. Hunt's No. 1 Jackson well, located in Choctaw County, Alabama, was accepted by the Jackson Oil Fraternity as a 5380 foot dry hole on Tuesday, February 8, 1944.
> Interest was revived and the wildcat returned to the active list by the time of the arrival on location of three truckloads of five inch OD casing on Tuesday morning, February 10th. (Dixie, February 17, 1944)

The well was completed February 16, 1944, as a small producer flowing 30 barrels of 19-degree gravity oil by heads. In spite of its marginal nature, it was Alabama's first oil discovery and created considerable excitement. Hunt went on to drill his No. 2 Robert Land, which found oil in the Eutaw formation and was completed pumping 50 barrels of 19-degree gravity oil per day plus some saltwater. The well was completed on August 24, 1944.

Hunt Oil Company erected a 1,200-barrel tank near the railroad at Gilbertown, Alabama, and built a four-car loading rack. A two-mile, four-inch line was laid from the first wells to the storage tank at Gilbertown.

In the early months of production, the crude was shipped to the Mexican Petroleum Corporation at Port Wentworth, Georgia. Soon Hunt extended the Gilbertown Field some 5.5 miles to the west with his No. 1 Scruggs well. A six-inch gathering system was constructed to pipe the oil from all wells in the field to the storage.

In 1945 the Carter Oil Company drilled on the block originally taken for it by the Vaugheys.

> About 1941 we took the Gilbertown Field for the Carter Oil Company in Alabama. This was the time when Germany was invading France and

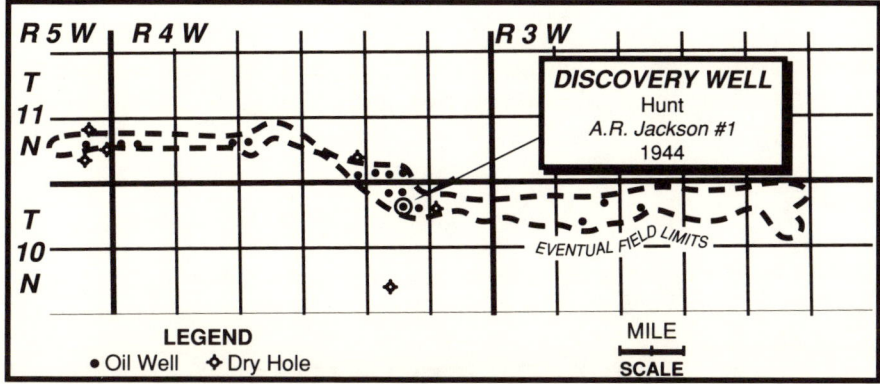

FIG. 16   Gilbertown and East Gilbertown Oil Fields, Choctaw County, 1945,
Alabama's first oil production

moving rather rapidly through that part of Europe and every morning before we went up to work we switched on the radio to see how far they had
gotten that day. France was collapsing rapidly. (E. A. Vaughey, March 11,
1987)

The company's first well, the No. 1 Alman, was completed from the Eutaw
with 50 feet of sand. This and subsequent wells drilled by Carter east of
Hunt's leases found the best part of the Gilbertown Field, known as East
Gilbertown.

After the pipeline had been extended to the Langsdale Field in Clarke
County, Mississippi, H. L. Hunt then let a contract for the building of
an asphalt refinery at Tuscaloosa, Alabama. The plant was to have a
3,500 barrel-a-day capacity, which would be ample for the production at
both Gilbertown and Langsdale. A pipeline was to be built to the Tombigbee River and the oil barged to Tuscaloosa.

By the end of 1945, 21 wells had been completed in the field, with a
cumulative production of 240,000 barrels.

The East Gilbertown Field, had four wells by the end of the year. Eventually, the Gilbertown producing area would be 13 miles long and have some
80 oil wells with a cumulative production approaching 14 million barrels
by 1992.

During this four-year period, 98 wells were drilled in Alabama, most of
them during the latter two years. By comparison, Mississippi had seen
900 wells drilled during the same period. Thus, 1945 ended with Alabama
having joined the ranks of oil producing states.

CHAPTER 41

Florida, 1942–1945

At the onset of World War II the states along the Atlantic Seaboard were consuming 40 percent of the petroleum being produced domestically. To supply the East Coast with petroleum, some 250 tankers shuttled continuously from ports in Texas and Louisiana, around the Florida peninsula, to ports along the Atlantic coastline.

These tankers immediately became prime targets for German submarines lying in wait along shipping lanes. By May 1942, some 55 tankers had been sunk (Williamson et al., 1963). Tanker losses continued through the first 18 months of the war. The beaches of Florida became coated with oil from the sunken tankers. Servicemen visiting the beaches reported "our bathing suits and skin became spotted with tar."

To relieve the situation, the "Big Inch" pipeline was completed by July 1943 enabling shipment of crude oil through an inland pipeline to the east coast. A few months later near the end of 1943 a second pipeline to transport refined petroleum products, the "Little Inch," was also completed. The beaches of Florida soon were clean by natural processes without help from man. No permanent ecological damage occurred.

The Oil Activity

Activity in Florida was very slow in 1942. During the entire year, there were only nine drilling operations. Three were dry and abandoned; four were shut down; and two were new locations. Leasing activity was generally concentrated on the Ocala Uplift. This huge uplift affected several counties in northern Florida and was centered near the town of Ocala.

FIG. 17   Sunnyland Field, Collier County, Florida, 1945, the state's first oil well

In Dade County, Blanchard had finally abandoned his No. 1 Everglades at 3,086 feet after failing to make a commercial gas well in the shallow zone and finding nothing deeper.

It appeared that 1943 would be another dismal period in the oil business of Florida. Only six wells were drilled during the year, but a significant event in October caught the attention of the oil companies.

Sunniland, Florida's First Oil Discovery

Near the southern tip of the Florida peninsula west of the Everglades, the Humble No. 1 Gulf Coast Realties Corporation well in Collier County had been staked in late 1942. Drilling was not started until mid-1943. (Gulf Coast Realties was one of the companies owned by Collier Company.) Humble found a show of oil in the well at 11,613 to 11,626 feet in a Lower Cretaceous reef. Pipe was set and an open-hole completion was attempted. The first oil swabbed on October 14, proved to be in the low 20-degree gravity range which was discouraging. Swabbing was continued for several days and the oil volume increased, but the well also made considerable saltwater. Nevertheless, from October 14 to October 23, 1943, a total of 1,350 barrels of oil were recovered:

> Wallace Pratt, Geologist, Standard Oil Company of New Jersey, has stated that the showing of oil in this Humble well is the most significant development in the past 30 years. He did not claim the well to be commercial, but called it a favorable showing.
>
> The state of Florida has a $50,000 reward for the first commercial well in the State. So far as is known, Humble has not applied for this reward yet. (Dixie, October 28, 1943)

The Sunniland Field, Florida's first discovery, was officially completed on November 26, 1943, pumping 140 barrels of oil per day with gravity of 20 degrees, plus 425 barrels of saltwater. The producing formation became known as the Sunniland Zone.

This set off a lease-buying spree of gigantic proportions:

> The largest lease play ever known in the southern states. . . . It is estimated that approximately six million acres were acquired in the last few months of the year [1943]. (Scouts, 1944)

> There are 67 counties in Florida and leases are being taken in *every* County. So far as is known, all leasing is just wildcat with no dope. The Companies who have vast holdings have held the prices to ten cents [bonus] and ten cents [rentals]. So far as is known, 25 cents is the top price that has been paid in Florida. Most all Companies have representatives in the State, with buying orders. (Dixie, November 18, 1943)

> In Dade County, Florida, W. G. Blanchard, et al have gone back into their No. 1 Everglades, which was temporarily abandoned, and are planning to make a deep test. The well has [already] reached a total depth of 7320 feet. (Dixie, November 18, 1943)

The Stampede to Florida

Companies rushed to open offices in Florida. Some of the newcomers were:

Company	Location	Representative	Title
Humble	Tampa	A. D. Hunter	Landman
		J. C. Cunningham	Landman
		Edward D. Prassler	Geologist
		W. A. Launey	Geologist
Magnolia	Tampa	Roy Jank	Landman (plus five lease brokers)

	Tallahassee	H. A. Selling	Geologist
Ohio	Tampa	Coes Mills	Geologist
Phillips	Tallahassee	Maurice Miesse	Landman
	Quincy	C. E. Moses	Geologist
Pure	Quincy	L. A. Finley	Landman
		Colonel Tibbets	Geologist
Shell	Tallahassee	W. W. Rand	Geologist
		Douglas Edmun	Scout
Sinclair	Tallahassee	O. G. Smith	Landman
		James L. Martin, Jr.	Geologist
	Waycross, Ga.	Vaughn Russum	Geologist
Stanolind	Tallahassee	Charles Shock	Manager
		Dale Chapman	Geologist
Sun	Tallahassee	D. J. Munroe	Geologist
		Dennis Chappin	Landman
	Lakeland	R. H. Weaver	Geological scout
	Waycross, Ga.	J. B. Halstead	Scout/Landman
Superior	————	George Lack	Landman
Tidewater	————	Monroe Bell	Landman
		J. L. Mackey	Landman
		Mr. Bush	Landman
Independents	Tallahassee	Leon Scanlon	
		Sidney Stubbs	
		Press Cochran	
		Merrill Harris	

The Florida Scout Association

In early 1944 a scout check was formed in Tallahassee. Until this time, the southeastern states had been handled by the Jackson Scout Association. Florida would be handled as a separate oil center, a distinction that Alabama would never have.

The 1945 National Oil Scouts and Landmen's Association yearbook indicates 15 active members in the Tallahassee Scout Association:

E. M. Cline (Shell), president
D. F. Callendar (Tide Water)
R. T. Chapman (Stanolind)
T. M. Crawford (Humble)

L. L. Curry (Texas Company)
M. A. Fromuth (Phillips)
L. B. Grayson (Superior)
Ray Janke (Mobile)
Murray Johnson (Gulf)
W. H. Mc Kinley (Stanolind)
Paul Mc Mahon (the California Company)
R. E. Morgan (Atlantic)
E. M. Pinckard (Pure)
O. G. Smith (Sinclair)
R. L. White (Sun)

A Dixie report of February 17, 1944, stated, "The organizational meeting of the proposed Florida Geological Society is to be held Thursday night at the Cherokee Hotel, Tallahassee, Florida." No city in Florida would become the recognized oil center, as Jackson had been in Mississippi. Tallahassee came the closest.

The $50,000 Reward

Apparently, later in 1944, Humble Oil & Refining Company applied to the state of Florida for the reward promised to the finder of the first commercial oil well in Florida. Humble's well had consistently pumped 125 barrels of oil, plus 400 to 500 barrels of water per day for several months. However, the state's representative ruled that Humble's well was not a commercial well after a 30-day production test, and Humble did not contest the decision. It was reversed a few months later.

The Dixie report of April 25, 1944, states that

> In Florida, the Humble Oil and Refining Company has been awarded the $50,000 award for discovering the first Commercial Oil production in the state. It is reported that Humble's officials have offered to split the reward with the University of Florida and the Florida State College for Women to be used as a scholarship fund. Humble's No. 1 Gulf Coast Reality Company has produced a total of approximately 10,000 barrels of oil.

The state legislature passed conservation laws for oil and gas in 1945 and implemented rules and regulations by 1946.

Sunniland Confirmation Well

A new depth record was set for Florida in 1944. The Humble No. 2 Gulf Coast Realty Company well reached a depth of 13,512 feet but failed to find oil in Sunniland and was abandoned.

The No. 4 Gulf Coast Realty was successful, however. On June 25, 1945, Humble officially completed this well as the second producing well in the Sunniland Field and in the state. The well pumped at the rate of 257 barrels of 19.2-degree gravity oil per day, plus 161 barrels of saltwater.

As 1945 came to a close, Humble was drilling additional wells in the Sunniland Field.

Florida's Oil Future

Florida had finally made the ranks of states with oil production. The Sunniland Field eventually produced 18 million barrels and paved the way for a number of more significant fields in the vicinity. Many problems had to be solved, not the least of which was getting the oil to market. Great marshlands separated the field from the coast where the crude could be transported by water to a refinery. Early transportation costs were so high that the field was barely profitable for the first decade.

Humble, which later became Exxon, remained the biggest producer in Florida into the 1990s.

CHAPTER 42

The Southeastern States
at the Close of 1945

President Harry Truman's use of two atomic bombs brought an ending to World War II much earlier than anticipated. The Japanese surrendered on August 15, 1945.

The Office of Petroleum Coordinator (OPC), which had become the Petroleum Administration of the War and had controlled the domestic oil industry for four years, stepped aside. Each state was to regulate its own oil and gas industry with no interference.

In August 1945 gasoline rationing and the 35-mile-per-hour speed limit were abolished. Some 25 million cars took to the roads in the United States.

Between December 1941 and August 1945 the U.S. oil industry had been called upon to produce and process more than six billion barrels of oil to supply military and domestic needs. Mississippi, Alabama, and Florida had joined the ranks of producing states. The long struggle of many companies and individuals was paying off.

By the close of 1945, Mississippi was a thriving oil-producing state with a number of significant discoveries having been made during the last seven years. A total of 1,746 wells had been drilled in Mississippi since the first one in 1911, of which 685 were oil wells, 171 gas wells, and 890 were dry holes. A total of 103,015,835 barrels of oil and 116,296,994 MCF of gas had been produced.

This placed Mississippi tenth in the nation as an oil-producing state in 1945. That year some 19 million barrels were produced, which amounted to one percent of the nation's total.

Southeastern States Production History as of December 31, 1945

MISSISSIPPI

Oil	1945	Cumulative as of December 31, 1945	
Baxterville	57,364	58,411	
Brookhaven	46,598	65,899	
Carthage Point	45,340	45,340	
Cary	10,637	52,945	
Cranfield	2,102,775	2,641,088	
Eucutta	2,059,703	2,515,698	
Fayette	23,939	23,939	
Flora	15,688	25,678	
Gwinville	106,909	109,553	
Heidelberg	2,897,747	4,338,939	
Hub	21,584	21,584	
Langsdale	93,159	93,159	
Mallallieu	116,845	149,148	
Pickens	2,072,604	6,606,838	
Quitman	2,100	2,100	
Soso	2,671	2,671	
Tinsley	9,332,425	86,180,938	
TOTAL	19,008,088	102,933,928	
Gas			
Amory	-0-	960,926	
Bruiensberg	24,253	33,827	(Shut-in)
Jackson	640,417	115,302,241	
TOTAL	664,670	116,296,994	

ALABAMA

	1945	Cumulative as of December 31, 1945
Gilbertown	184,210	243,308
East Gilbertown	1,889	1,889
TOTAL	186,099	245,197

FLORIDA

Sunniland	27,400	43,254

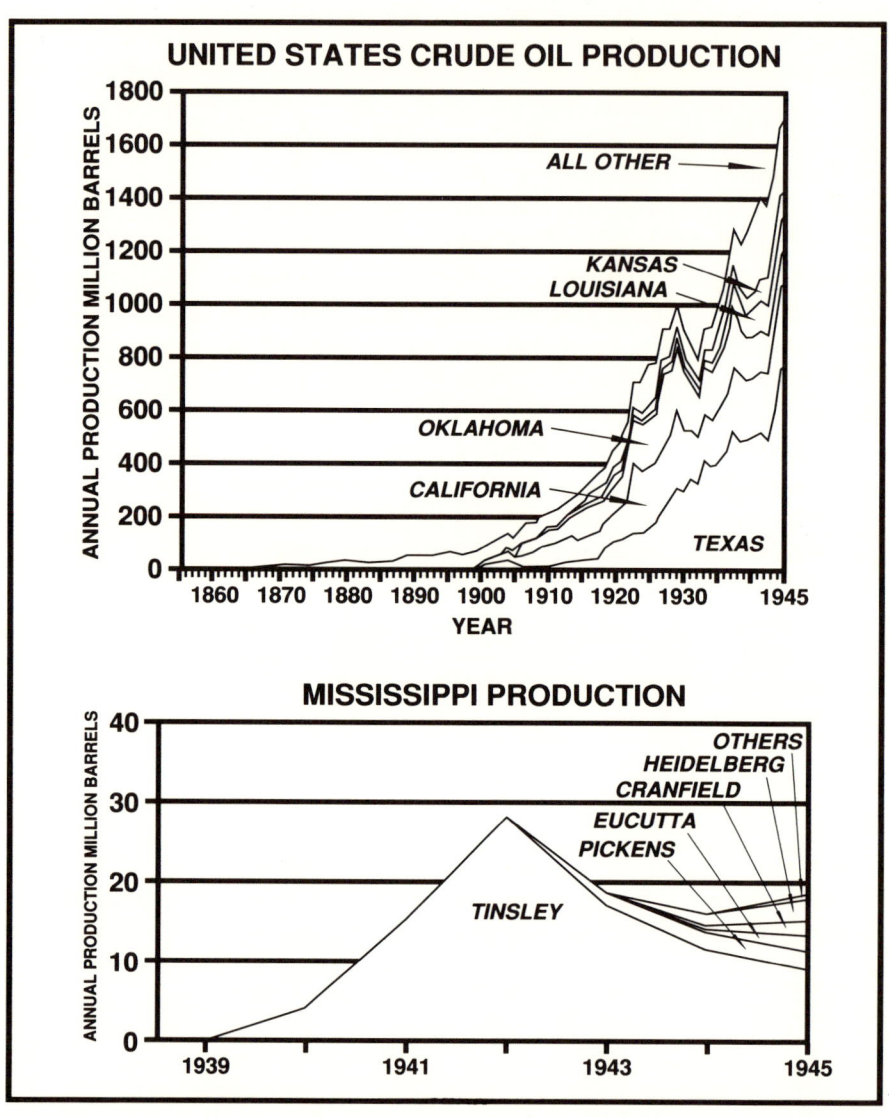

FIG. 18   Production listed as "all other" includes Pennsylvania, New York, Illinois, Wyoming, Arkansas, Ohio, New Mexico, West Virginia, Michigan, Kentucky, Tennessee, Indiana, Montana, Mississippi, Colorado, Nebraska, Missouri, Alabama, Florida, and Virginia.

FIG. 18A   Production listed as "others" include Baxterville, Brookhaven, Carthage Point, Cary, Fayette, Flora, Gwinville, Hub, Langsdale, Mallalieu, Quitman, and Soso.

The fields discovered in Mississippi from 1939 to 1945 had produced over a billion barrels of oil by 1990, representing 50 percent of the oil produced in the state to that date. The gas fields discovered during this period had produced 3.5 trillion cubic feet of gas by 1990, representing half the gas produced in Mississippi to 1990.

Alabama and Florida had small amounts of production by 1945, but both states would reach their zenith many years in the future. By the end of 1945, Alabama had 283 wells drilled, according to the state records, of which 15 produced oil, along with an estimated 40 early shallow gas wells. The remainder were dry holes. Florida was the least explored of the three states, with 91 wells having been drilled finding two oil producers, no gas wells, and 89 dry holes.

At the end of World War II, the need for petroleum products was switching from military to civilian uses. New automobiles were in great demand. There was a worldwide shortage of petroleum. The future of petroleum exploration in the southeastern states looked very bright.

References

Blakey, 1985
Blakey, Ellen Sue. *Oil on Their Shoes, Petroleum Geology to 1918*. Tulsa: American Association of Petroleum Geology, 1985.

Brinson, 1984
Brinson, Carroll. *Always a Challenge, MP&L First Sixty Years*. Jackson: Oakdale Press, 1984.

Brunini, 1990
Brunini, Edward. Letter and files from hearings concerning *Tinsley Crude Transportation, 1940–1943*, 1990.

Burns, 1990
Burns, Henry Burns, Sr. Interview with author, Jackson, Mississippi, 1990.

Castle, 1988
Castle, Craig. Interview with author. Jackson, Mississippi, 22 January 1988.

Clark, 1952
Clark, James A. and Halbouty, Michel T. *Spindletop*. New York: Random House, 1952.

Clark, 1955
Clark, J. Stanley. *The Oil Century, from the Drake Well to the Conservation Era*. Norman, Oklahoma: University of Oklahoma Press, 1955.

Clark, 1963
Clark, James A. *The Chronological History of the Petroleum and Natural Gas Industries*. Houston, Texas: Clark Book Company, 1963.

Clark and Halbouty, 1972
Clark, James A. and Halbouty, Michel T. *The Last Boom*. New York: Random House Press, 1972.

Dalrymple, 1989
Dalrymple, Arch. Letter to author, 1989.

Davis and Lambert, 1963
Davis, David C. and Lambert, Ernest H., Jr. eds. *Mesozoic-Paleozoic Producing Areas of Mississippi & Alabama, Volume II.* Jackson, Mississippi, Mississippi Geological Society, 1963.

Davis, 1964
Davis, Ralph E. *The Natural Gas Industry.* James A. Clark, ed. Ralph E. Davis Publisher, 1964.

DeGolyer and MacNaughton, 1990
DeGolyer, E. L. and L. W. MacNaughton. *Twentieth Century 1990, Petroleum Statistics.* Dallas, 1990.

Dixie, 1941–1945
Dixie Geological Survey Reports 1941 through 1945.

Evans, 1988
Evans, James. Interview with author. Jackson, Mississippi, 20 December 1988.

Fanning, 1936
Fanning, Leonard M. *The Rise of American Oil.* New York and London: Harper & Brothers Publishers, 1936 and 1948.

Franks, 1982
Franks, Kenny A. and Lambert, Paul F. *Early Louisiana and Arkansas Oil 1901– 1946.* College Station, Texas: Texas A & M Press, 1982.

Frascogna, 1957
Frascogna, Xavier M. ed. *Mesozoic-Paleozoic Producing Areas of Mississippi and Alabama.* Jackson, Mississippi: Mississippi Geological Society, 1957.

Green, 1987
Green, Gardiner, Sr. Interview with author. Laurel, Mississippi, 18 March 1987.

Guffey, 1989
Guffey, Roy. Interview with author. Jackson, Mississippi, 27 November 1989.

Hayes, 1989
Hayes, Charles. Interview with author. Jackson, Mississippi, 18 November 1989.

Hearin, 1988
Hearin, Robert. Interview with author. Jackson, Mississippi, 13 September 1988.

Henderson and Shaw, 1945
Henderson, Harry and Shaw, Sam. *Collier Magazine, Ole Miss Strikes It Rich.* 10 February 1945.

Hester, 1989
Hester, W. E., Jr. Conversations and correspondence with author. Jackson, Mississippi, 9 August 1989.

Hobson, 1920
Hobson, S. A. *Ups and Downs in Old Alabama, From the Oil and Gas Standpoint, A Story in Three Chapters: Then, Some Later, and Now.* Dallas: Geological Library, Southern Methodist University, 1920.

Hopkins, 1916
Hopkins. *Structure of the Vicksburg-Jackson Area, Mississippi with Special Reference to Oil and Gas.* United States Geological Survey, Bulletin 641-D, 18 July 1916.

Hughes, 1986
Hughes, J. Willis. Interview with author. Jackson, Mississippi, 26 September 1986.

International, 1980
International Oil & Gas, Development Yearbook, 1980, Review of 1978–1979, Part II, Vol. XLIX.

Ickes, 1943
Ickes, Harold. *Fightin Oil.* Reprinted by permission of the publisher Alfred A. Knopf, New York: Ryerson Press, 1943.

Ivey, 1967
Ivey, W. T. *Southern Natural Gas System History.* Birmingham: March 1, 1967.

Jeffreys, 1989
Jeffreys, Geoffrey. Interview with author. Jackson, Mississippi, 16 November 1989.

Kelly, 1990
Kelly, George. Conversations with author. (Kelly is a retired MP&L employee who was a boy in Cleveland at this time.) August, 1990.

Knight, 1987
Knight, Wilbur. Interview with author. Jackson, Mississippi, 24 July 1987.

Ladner, 1989
Hilton Ladner. Interview with author. Jackson, Mississippi, 7 August 1989.

Lambdin, 1990
Lambdin, Harry. Informal interviews with author. Jackson, Mississippi, 1987– 1990.

Larson and Porter, 1959
Larson, Henrietta M. and Porter, Kenneth Wiggins. *History of Humble Oil &*

Refining Company. Business History Foundation. New York: Harper Collins, 1959.

Lloyd, 1989
Lloyd, Baldwin. Informal interviews with author. Jackson, Mississippi, 1989.

Lowe, 1925
Lowe, E. N. Mississippi State Geological Survey, Bulletin 20. *Geology and Mineral Resources of Mississippi,* 1925.

Lynn, Date Not Available
Lynn, Homer. Interview with author. Jackson, Mississippi, date not available.

Mayeaux, 1991
Mayeaux, Joseph. Informal interview with author. Jackson, Mississippi, 1991.

Mellen, 1987
Mellen, Fred. Interview with author. Jackson, Mississippi, 5 March 1987.

Mellen, 1952
Mellen, Frederic F. *Oil & Gas Possibilities, Black Warrior Basin, Alabama and Mississippi.* 15 June 1952.

Mellen, 1940
Mellen, Frederic F. Mississippi State Geological Survey, Bulletin 39. *Yazoo County Mineral Resources,* 1940.

Miller, 1990
Miller, David. Letter to author dated 22 August 1990. Deed dated 28 November 1916, Recorded in Book DE, Page 78, Yazoo County, Mississippi. Deed dated 10 January 1934, Recorded in Book GS, Page 282, 1990.

Monroe, 1937
Monroe, Watson Hiner and Toler, Henry Niles. *Mississippi State Geological Survey, Bulletin 36.* The Jackson Gas Field and The State Deep Test Well, 1937.

Morgan, 1987
Morgan, Jimmy. Interview with author. Laurel, Mississippi, 18 March 1987.

Morgan, 1990
Morgan, Dan, and Morgan, Martha Mrs. Interview with author. Jackson, Mississippi, 1990.

Norman, 1988
Norman, Bud. Interview with author. Jackson, Mississippi, 27 October 1988.

Oil Trade Journal, Inc., 1923
Oil Trade Journal, Inc. *The Petroleum Register, Standard Directory and Statistical Records of the Petroleum Industry,* 1923.

Owen, 1975

Owen, Edgar Wesley. *Trek of the Oil Finders: A History of Exploration for Petroleum.* Tulsa: American Association of Petroleum Geologists, 1975.

Pepper, 1989
Pepper, Frances. Interview with author. Jackson, Mississippi, 18 July 1989.

Petroleum, 1984–86
Petroleum Independent Industry in Your State (1984) (1985) and (1986).

Phillips, 1989
Phillips, Robert R. Letters dated 15 December 1989, 18 December 1989.

Pittman, 1988
Pittman, Crymes. Interview with author. Jackson, Mississippi, 8 November 1988.

Reese, 1986
Reese, Don. Interview with author. Jackson, Mississippi, 30 June 1986.

Ridgway, 1989
Ridgway, Julius. Interview with author. Jackson, Mississippi, 28 June 1989.

Ridgway, 1989
Ridgway, Robert and Bryant. Interview with author. Jackson, Missisippi, 13 October 1989.

Schmidt, 1990
Schmidt, Walter. Florida Geological Survey, Bureau of Geology, Florida Department of Natural Resources. *History of the Florida Geological Survey.* Tallahassee, Florida, 1990.

Scouts, 1930
National Oil Scouts Association of America. "Oil and Gas Developments." Annual Yearbook 1930 through 1945.

Smith, 1990
Smith, Frank. *Hilgard's Studies Still Influence State Agriculture.* Jackson: Jackson *Clarion Ledger,* 15 July 1990, Page 3F.

Steffey, 1928–1941
Steffey Scout Reports, 1928 through 1941.

Storey, 1987–1990
Storey, J. B. Interviews with author. New Orleans, Louisiana, 1987 and 1990.

Stotz and Jamison, 1938
Stotz, Lewis. *History of the Gas Industry.* New York: Press of Stettiner Bros., 1938.

Tinkle, 1970
Tinkle, Lon. *Mr. De, A Biography of Everette Lee DeGolyer,* Boston and Toronto. Canada: Little, Brown, & Company, 1970.

Toler, 1987

Toler, Ruth. Interview with author. Jackson, Mississippi, 23 June 1987.

Vaughey, 1987

Vaughey, Emmett and Bill Vaughey. Interview with author. Jackson, Mississippi, 11 March 1987.

Vestal, 1943

Vestal, Franklin Earl. *Mississippi State Geological Survey, Bulletin 57, Mineral Resources.* University, Mississippi, 1943.

Who's Who, 1952. *Who's Who in the South and Southwest.* Chicago: A. N. Marquis, 1952.

Williams, 1989

Williams, Christine (Granddaughter of Big Boy Love). Interview with author. 2 August 1989.

Williamson and Daum, 1959

Williamson, Harold F., and Arnold R. Daum. *The American Petroleum Industry, Volume I 1859–1899.* Northwestern University Press, 1959.

Williamson, et al. 1963

Williamson, Harold F., Ralph L. Andreno, Arnold R. Daum. *The American Petroleum Industry.* Northwestern University Press, 1963.

Womack, 1987

Womack, Robert. Interview with author. Jackson, Mississippi, 16 April 1987.

Yergin, 1991

Yergin, Daniel. *The Prize, The Epic Quest for Oil, Money, and Power.* New York: Simon and Schuster, 1991.

Index